ZigBee Wireless Networks and Transceivers

T0146332

ZigBee Wireless Networks and Transceivers

Shahin Farahani

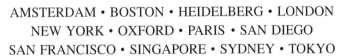

AMSTERDAM • BOSTON • HEIDELBERG • LONDON
NEW YORK • OXFORD • PARIS • SAN DIEGO
SAN FRANCISCO • SINGAPORE • SYDNEY • TOKYO

Newnes is an imprint of Elsevier

ELSEVIER

Newnes

Newnes is an imprint of Elsevier
30 Corporate Drive, Suite 400, Burlington, MA 01803, USA
Linacre House, Jordan Hill, Oxford OX2 8DP, UK

Library of Congress Cataloging-in-Publication Data
Application submitted.

British Library Cataloguing-in-Publication Data
A catalogue record for this book is available from the British Library.

ISBN: 978-0-7506-8393-7

For information on all Newnes publications
visit our Web site at: www.books.elsevier.com

08 09 10 11 12 13   10 9 8 7 6 5 4 3 2 1

Printed in the United States of America.

To my family

Contents

Foreword

ZigBee® Wireless Technology is the leading global standard for implementing low-cost, low-data-rate, short-range wireless networks with extended battery life. The ZigBee Alliance is an association of companies working together to enable reliable, cost-effective, low-power, wirelessly networked monitoring and control products based on an open global standard. The specification documents released by the ZigBee Alliance and IEEE 802.15 WPAN Working Group are the official sources for implementing ZigBee and IEEE 802.15.4-based wireless networks. However, these lengthy documents can be overwhelming for people with time constraints or technical backgrounds that are more product-focused. Therefore, there is a need for a comprehensive resource that not only contains the in-depth technical information but also provides high-level overviews of the fundamentals of ZigBee wireless networking. This book, by Dr. Shahin Farahani, addresses this need with a concise tutorial on ZigBee technology.

This book provides a complete picture of ZigBee wireless networking, from radio frequency (RF) and Physical layer (PHY) considerations up to Application layer development. The book is very well organized and the materials are easy to follow, with well-balanced coverage of the protocol. In addition, Chapter 3 of this book contains the frequently used IEEE 802.15.4 and ZigBee tables and frame formats and can act as a quick reference for these standards.

This book serves as an introduction to IEEE 802.15.4, the ZigBee standard and its applications. Just as important, the content is sufficiently thorough to provide deep understanding of the practical considerations of implementing any size ZigBee wireless network. For example, one of the more interesting potentials for the technology within the "intelligent RFID" space is to allow the ability to estimate the location of objects or personnel. Chapter 7 of this book is a good introduction to various approaches that can be used in location estimation and practical limitations of each method. The book is accompanied by a Website that contains extremely valuable calculators for range estimation and battery life analysis in various practical use-case scenarios.

The author has been a ZigBee systems engineer for several years, giving him a broad understanding of ZigBee and IEEE 802.15.4 wireless networking. Dr. Farahani's in-depth technical knowledge and communication skills deliver a book that is a must-read for engineers involved in short-range, low-data-rate wireless networking. Although the book is designed for readers of various backgrounds in the industry, from wireless system designers to middleware engineers and end-product OEM developers, the majority of its content can be thought of as part of an academic course dedicated to wireless sensor networks or short-range, low-data-rate wireless networking.

In short, I recommend this book to system designers, technical managers, wireless engineers, and researchers who are interested in gaining a comprehensive understanding of the principals and applications of ZigBee for low-power, low-data-rate wireless networking.

Dr. Robert F. Heile
Chair, IEEE 802.15
Chair, ZigBee Alliance

Preface

This book was written using a tutorial approach and is intended to appeal to a broad audience. The book covers the fundamentals of short-range wireless networking using ZigBee™ and IEEE 802.15.4™ standards. The ZigBee and IEEE 802.15.4 standards are covered with the same level of detail. In addition to technical details, the book contains high-level overviews that would be informative for a reader who is only interested in understanding the general concept of ZigBee wireless networking. This book provides the big picture of ZigBee wireless networking, from the radio frequency (RF) and physical layer considerations up to the application layer details. Considering the multidisciplinary nature of this book, the required background materials are also provided.

This book is expected to be useful for engineers and technical managers who want to know the principles and applications of ZigBee in short-range wireless networking. The reader should be able to follow most of the materials in this book as long as he or she has some basic knowledge in electrical engineering and wireless networking. The book is highly illustrated and contains 200 figures and tables. Although the book is designed for self-learning individuals in the industry, the majority of its content can be taught as part of a course dedicated to short-range, low-data-rate wireless sensor networking as well.

Overview of the Content

Chapter 1 is an introduction to short-range wireless networking using ZigBee and IEEE 802.15.4. This chapter summarizes many of the topics that are discussed in further detail in the reminder of this book. Chapter 1 starts with describing the ZigBee standard and the ways it differs from other existing short-range wireless networking standards. Then the fundamentals of the ZigBee and IEEE 802.15.4 standard are reviewed. Chapter 2 is brief and provides a number of ZigBee wireless networking examples, including home automation, consumer electronics, and healthcare.

Chapter 3 is a detailed description of IEEE 802.15.4 and ZigBee protocol layers, including security features. This chapter acts as a summary of the IEEE 802.15.4 and ZigBee standard documents. Chapter 3 provides information regarding the 2007 edition of the ZigBee standard, whenever needed, along with the 2006 edition. Appendix D summarizes some of the differences between ZigBee-Pro (2007) and ZigBee-2006. If the level of detail in Chapter 3 exceeds your interest or background, you can skip that chapter and continue with Chapter 4 as long as you have read Chapter 1.

Basic information regarding transceivers developed for the IEEE 802.15.4 standard is provided in Chapter 4. An IEEE 802.15.4-compliant transceiver must meet specific requirements outlined in the standard itself. This chapter helps in understanding these requirements and in comparing and contrasting performance of various transceivers. In a wireless sensor network (WSN), the sensor output is an analog signal and must be converted to digital. The basics of analog-to-digital to converters (ADCs) and their performance metrics are provided in Chapter 4.

Chapter 5 studies the effect of environment on a ZigBee wireless network. In reality, the performance of a wireless network can be affected by the environment. Chapter 5 provides a number of techniques that can be used to mitigate performance degradation due to environment. Range estimation and the methods that can be used to improve the range are also covered in Chapter 5. The antenna can be designed specifically for each application scenario and may greatly affect the performance of a wireless node. Fundamental properties of antennas and some of the basic antenna shapes used in short-range wireless networks are provided in Chapter 5. Regulatory requirements in North America, Europe, and Japan are also reviewed in this chapter.

A battery-powered node developed for ZigBee wireless networking may operate for several years before its battery must be replaced. The expected battery life depends on the application scenario, and Chapter 6 provides a simple way of calculating the estimated battery life. Battery life is not only a function of hardware-level performance of each node but also depends on the operation efficiency of the network. Chapter 6 reviews both hardware- and network-level battery life extension methods.

One of the applications of short-range wireless networking is estimating the location of moving objects. Chapter 7 provides a number of methods that can be used for location estimation using a ZigBee network. In contrast to a global positioning system (GPS), the short-range location-estimation methods rely solely on the information gathered locally from other nearby nodes to estimate the location of a moving object.

ZigBee networks operate in an unlicensed band, shared by many other wireless networks such as WiFi™ for wireless Internet access. Chapter 8 provides an introduction to a coexistence mechanism and reviews some of the methods used in IEEE 802.15.4 wireless networking to increase the robustness of the network to interferences. Chapter 9 reviews some of alternative short-range wireless networking standards, including IPv6 over IEEE 802.15.4 and WirelessHART™. This book contains five appendices. The summary of the differences between ZigBee-Pro/2007 and ZigBee-2006 is provided in appendix D. Appendix E reviews the building blocks inside a transceiver using a tutorial approach. Appendix E is not a prerequisite for Chapter 4.

Supplements

This book comes with a dedicated Website (www.LearnZigBee.com) instead of a companion CD. The applications of short-range wireless networking are growing and the associated standards and regulations may change over time. A dedicated Website for the book allows us to provide up-to-date information to readers even after the book is published. You can also post questions or comments on the LearnZigBee.com Website. The content of the Website includes, but is not limited to, the following:

- A simple spreadsheet (calculator) for range and battery life analysis

- Links to useful tools and additional materials

- Links to standard documents, including the latest versions of IEEE 802.15.4 and ZigBee standards

- Links to regulatory requirements documents, including FCC and ETSI

- Links to related datasheets (e.g., IEEE 802.15.4 transceivers)

- Additional references for short-range wireless networking

The simple spreadsheet (calculator) we mentioned was created based on the materials discussed in this book. For example, you can use this spreadsheet to calculate the expected battery life of a node or estimate the range improvement if an external power amplifier (PA) is added to a node. Although the spreadsheet is simple and self-explanatory, a document is posted on the Website, explaining the calculators and walking you through a number of examples.

Acknowledgments

I would like to thank Dr. Darioush Keyvani, Phil Beecher, and Louis E. Frenzel for providing detailed reviews of the book. The book's content was put together little by little over the past several years based on my experiences as a ZigBee system architect at Freescale Semiconductor Inc. This book would not be possible without the work of a talented, knowledgeable, and diverse team. Therefore, I would like to thank my colleagues at Freescale for their help and continued support, in particular Clinton Powell, Brett Black, Jon Adams, and Dr. Kuor-Hsin Chang. I am especially grateful to my wife, Nazanin Darbanian, who not only provided valuable technical feedback but also was patient and supportive during the writing of this book. I would also like to thank the staff at Elsevier Science, particularly Harry Helms, Melinda Ritchie, Rachel Roumeliotis, and Heather Scherer, for their support.

Abbreviations

Abbreviation	Phrase
AIB	Application Support Layer Information Base
AES	Advanced Encryption Standard
AF	Application Framework
APDU	Application Support Sublayer Protocol Data Unit
APL	Application Layer
APS	Application Support Sublayer
APSDE	Application Support Sublayer Data Entity
APSDE-SAP	APSDE-Service Access Point
APSME	Application Support Sublayer Management Entity
APSME-SAP	APSME-Service Access Point
ASDU	APS Service Data Unit
ASK	Amplitude Shift Keying
AWGN	Additive White Gaussian Noise
BPSK	Binary Phase-Shift Keying
BRT	Broadcast Retry Timer
BSN	Beacon Sequence Number
BTR	Broadcast Transaction Record
BTT	Broadcast Transaction Table
CAP	Contention Access Period
CBC-MAC	Cipher Block Chaining Message Authentication Code
CCA	Clear Channel Assessment
CCM	Counter with CBC-MAC
CCM*	CCM (extended version)
CFP	Contention-Free Period

CRC	Cyclic Redundancy Check
CSMA-CA	Carrier Sense Multiple Access with Collision Avoidance
DSN	Data Sequence Number
DSSS	Direct Sequence Spread Spectrum
ED	Energy Detection
EIRP	Effective Isotropic Radiated Power
ERP	Effective Radiated Power
EVM	Error Vector Magnitude
FCS	Frame Check Sequence
FFD	Full-Function Device
FH	Frequency Hopping
FHSS	Frequency Hopping Spread Spectrum
GTS	Guaranteed Time Slot
HDR	Header
IB	Information Base
IFS	Interframe Spacing
ISM	Industrial, Scientific, and Medical
LIFS	Long Interframe Spacing
LQI	Link Quality Indicator
LR-WPAN	Low-Rate Wireless Personal Area Network
LSB	Least Significant Bit
MAC	Medium Access Control
MCPS	MAC Common Part Sublayer
MCPS-SAP	MAC Common Part Sublayer Service Access Point
MFR	MAC Footer
MHR	MAC Header
MIC	Message Integrity Code
MLME	MAC Layer Management Entity
MLME-SAP	MAC Layer Management Entity Service Access Point
MSB	Most Significant Bit
MPDU	MAC Protocol Data Unit
MSDU	MAC Service Data Unit
NF	Noise Figure
NHLE	Next Higher Layer Entity
NIB	Network Layer Information Base
NLDE	Network Layer Data Entity
NLDE-SAP	Network Layer Data Entity Service Access Point

NLME	Network Layer Management Entity
NLME-SAP	Network Layer Management Entity Service Access Point
NPDU	Network Layer Protocol Data Unit
NSDU	Network Service Data Unit
NWK	Network Layer
OCDM	Orthogonal Code Division Multiplexing
O-QPSK	Offset Quadrature Phase-Shift Keying
OSI	Open Systems Interconnection
PAN	Personal Area Network
PC	Personal Computer
PD	PHY Data
PD-SAP	PHY Data Service Access Point
PER	Packet Error Rate
PHR	PHY Header
PHY	Physical Layer
PIB	PAN Information Base
PLME	Physical Layer Management Entity
PLME-SAP	Physical Layer Management Entity Service Access Point
PN	Pseudorandom Noise
POS	Personal Operating Space
PPDU	PHY Protocol Data Unit
PSD	Power Spectral Density
PSDU	PHY Service Data Unit
PSSS	Parallel Sequence Spread Spectrum
QOS	Quality of Service
RF	Radio Frequency
RFD	Reduced Function Device
RREP	Route Reply
RREQ	Route Request
RX	Receiver
SAP	Service Access Point
SFD	Start-of-Frame Delimiter
SHR	Synchronization Header
SIFS	Short Interframe Spacing
SIR	Signal-to-Interference Ratio
SNR	Signal-to-Noise Ratio
SKKE	Symmetric-Key Key Establishment

SSP	Security Services Provider
TRX	Transceiver
TX	Transmitter
WLAN	Wireless Local Area Network
WPAN	Wireless Personal Area Network
ZDO	ZigBee Device Object

CHAPTER 1

ZigBee Basics

This chapter is an introduction to the ZigBee standard for short-range wireless networking. The goal of this chapter is to provide a brief overview of ZigBee's fundamental properties, including its networking topologies, channel access mechanism, and the role of each protocol layer. The topics discussed in this chapter are covered in more detail in the reminder of this book.

1.1   What Is ZigBee?

ZigBee is a standard that defines a set of communication protocols for low-data-rate short-range wireless networking [1]. ZigBee-based wireless devices operate in 868 MHz, 915 MHz, and 2.4 GHz frequency bands. The maximum data rate is 250 K bits per second. ZigBee is targeted mainly for battery-powered applications where low data rate, low cost, and long battery life are main requirements. In many ZigBee applications, the total time the wireless device is engaged in any type of activity is very limited; the device spends most of its time in a power-saving mode, also known as *sleep mode*. As a result, ZigBee-enabled devices are capable of being operational for several years before their batteries need to be replaced.

One application of ZigBee is in-home patient monitoring. A patient's blood pressure and heart rate, for example, can be measured by wearable devices. The patient wears a ZigBee device that interfaces with a sensor that gathers health-related information such as blood pressure on a periodic basis. Then the data is wirelessly transmitted to a local server, such as a personal computer inside the patient's home, where initial analysis is performed. Finally, the vital information is sent to the patient's nurse or physician via the Internet for further analysis [2].

Another example of a ZigBee application is monitoring the structural health of large-scale buildings [3]. In this application, several ZigBee-enabled wireless sensors (e.g., accelerometers) can be installed in a building, and all these sensors can form a single wireless network to gather the information that will be used to evaluate the building's structural health and detect signs of possible damage. After an earthquake, for example, a building could require inspection before it reopens to the public. The data gathered by the sensors could help expedite and reduce the cost of the inspection. A number of other ZigBee application examples are provided in Chapter 2.

The ZigBee standard is developed by the ZigBee Alliance [4], which has hundreds of member companies, from the semiconductor industry and software developers to original equipment manufacturers (OEMs) and installers. The ZigBee Alliance was formed in 2002 as a nonprofit organization open to everyone who wants to join. The ZigBee standard has adopted IEEE 802.15.4 as its Physical Layer (PHY) and Medium Access Control (MAC) protocols [5]. Therefore, a ZigBee-compliant device is compliant with the IEEE 802.15.4 standard as well.

The concept of using wireless communication to gather information or perform certain control tasks inside a house or a factory is not new. There are several standards, reviewed in Chapter 9, for short-range wireless networking, including IEEE 802.11 Wireless Local Area Network (WLAN) and Bluetooth. Each of these standards has its advantages in particular applications. The ZigBee standard is specifically developed to address the need for very low-cost implementation of low-data-rate wireless networks with ultra-low power consumption.

The ZigBee standard helps reduce the implementation cost by simplifying the communication protocols and reducing the data rate. The minimum requirements to meet ZigBee and IEEE 802.15.4 specifications are relatively relaxed compared to other standards such as IEEE 802.11, which reduces the complexity and cost of implementing ZigBee compliant transceivers.

The duty cycle is the ratio of the time the device is active to the total time. For example, if a device wakes up every minute and stays active for 60 ms, then the duty cycle of this device is 0.001, or 0.1%. In many ZigBee applications, the devices have duty cycles of less than 1% to ensure years of battery life.

1.2   ZigBee versus Bluetooth and IEEE 802.11

Comparing the ZigBee standard with Bluetooth and IEEE 802.11 WLAN helps us understand how ZigBee differentiates itself from existing established standards. (A more comprehensive comparison is provided in Chapter 9.) Figure 1.1 summarizes the basic characteristics of these three standards.

IEEE 802.11 is a family of standards; IEEE 802.11b is selected here because it operates in 2.4 GHz band, which is common with Bluetooth and ZigBee. IEEE 802.11b has a high data rate (up to 11 Mbps), and providing a wireless Internet connection is one of its typical applications. The indoor range of IEEE 802.11b is typically between 30 and 100 meters. Bluetooth, on the other hand, has a lower data rate (less than 3 Mbps) and its indoor range is typically 2–10 meters. One popular application of Bluetooth is in wireless headsets, where Bluetooth provides the means for communication between a mobile phone and a hands-free headset. ZigBee has the lowest data rate and complexity among these three standards and provides significantly longer battery life.

ZigBee's very low data rate means that it is not the best choice for implementing a wireless Internet connection or a CD-quality wireless headset where more than 1 Mbps is desired. However, if the goal of wireless communication is to transmit and receive simple commands and/or gather information from sensors such as temperature or humidity sensors, ZigBee provides the most power and the most cost-efficient solution compared to Bluetooth and IEEE 802.11b.

1.3   Short-Range Wireless Networking Classes

Short-range wireless networking methods are divided into two main categories: wireless local area networks (WLANs) and wireless personal area networks (WPANs).

WLAN is a replacement or extension for wired local area networks (LANs) such as Ethernet (IEEE 802.3). A WLAN device can be integrated with a wired LAN network,

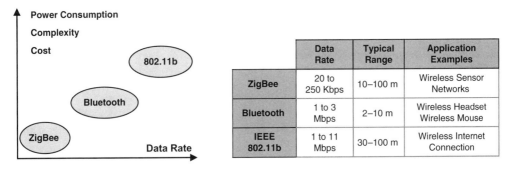

	Data Rate	Typical Range	Application Examples
ZigBee	20 to 250 Kbps	10–100 m	Wireless Sensor Networks
Bluetooth	1 to 3 Mbps	2–10 m	Wireless Headset Wireless Mouse
IEEE 802.11b	1 to 11 Mbps	30–100 m	Wireless Internet Connection

Figure 1.1: Comparing the ZigBee Standard with Bluetooth and IEEE 802.11b

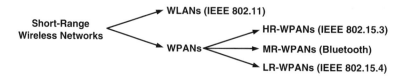

Figure 1.2: Short-range Wireless Networking Classes

and once the WLAN device becomes part of the network, the network treats the wireless device the same as any other wired device within the network [6]. The goal of a WLAN is to maximize the range and data rate.

WPANs, in contrast, are not developed to replace any existing wired LANs. WPANs are created to provide the means for power-efficient wireless communication within the personal operating space (POS) without the need for any infrastructure. POS is the spherical region that surrounds a wireless device and has a radius of 10 meters (33 feet) [5].

WPANs are divided into three classes (see Figure 1.2): high-rate (HR) WPANs, medium-rate (MR) WPANs, and low-rate (LR) WPANs [7]. An example of an HR-WPAN is IEEE 802.15.3 with a data rate of 11 to 55 Mbps [8]. This high data rate helps in applications such as real-time wireless video transmission from a camera to a nearby TV. Bluetooth, with a data rate of 1 to 3Mbps, is an example of an MR-WLAN and can be used in high-quality voice transmission in wireless headsets. ZigBee, with a maximum data rate of 250Kbps, is classified as an LR-WPAN.

1.4   The Relationship Between ZigBee and IEEE 802.15.4 Standards

One of the common ways to establish a communication network (wired or wireless) is to use the concept of *networking layers*. Each layer is responsible for certain functions in the network. The layers normally pass data and commands only to the layers directly above and below them.

ZigBee wireless networking protocol layers are shown in Figure 1.3. ZigBee protocol layers are based on the Open System Interconnect (OSI) basic reference model [9]. Dividing a network protocol into layers has a number of advantages. For example, if the protocol changes over time, it is easier to replace or modify the layer that is affected by the change rather than replacing the entire protocol. Also, in developing an application, the lower layers of the protocol are independent of the application and can be obtained from a

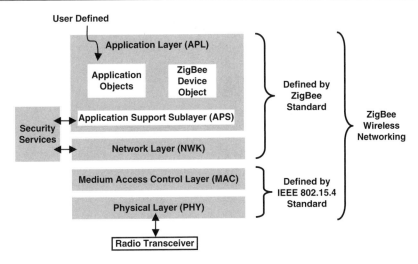

Figure 1.3: ZigBee Wireless Networking Protocol Layers

third party, so all that needs to be done is to make changes in the application layer of the protocol. The software implementation of a protocol is known as *protocol stack software*.

As shown in Figure 1.3, the bottom two networking layers are defined by the IEEE 802.15.4 standard [5]. This standard is developed by the IEEE 802 standards committee and was initially released in 2003. IEEE 802.15.4 defines the specifications for PHY and MAC layers of wireless networking, but it does not specify any requirements for higher networking layers.

The ZigBee standard defines only the networking, application, and security layers of the protocol and adopts IEEE 802.15.4 PHY and MAC layers as part of the ZigBee networking protocol. Therefore, any ZigBee-compliant device conforms to IEEE 802.15.4 as well.

IEEE 802.15.4 was developed independently of the ZigBee standard, and it is possible to build short-range wireless networking based solely on IEEE 802.15.4 and not implement ZigBee-specific layers. In this case, the users develop their own networking/application layer protocol on top of IEEE 802.15.4 PHY and MAC (see Figure 1.4). These custom networking/application layers are normally simpler than the ZigBee protocol layers and are targeted for specific applications.

One advantage of custom proprietary networking/application layers is the smaller size memory footprint required to implement the entire protocol, which can result

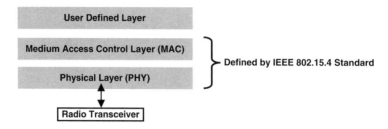

Figure 1.4: A Networking Protocol can be Based on IEEE 802.15.4 and not Conform to the ZigBee Standard

in a reduction in cost. However, implementing the full ZigBee protocol ensures interoperability with other vendors' wireless solutions and additional reliability due to the mesh networking capability supported in ZigBee. The decision of whether or not to implement the entire ZigBee protocol or just IEEE 802.15.4 PHY and MAC layers depends on the application and the long-term plan for the product.

Physical-level characteristics of the network are determined by the PHY layer specification; therefore, parameters such as frequencies of operation, data rate, receiver sensitivity requirements, and device types are specified in the IEEE 802.15.4 standard.

This book covers the IEEE 802.15.4 standard layers and the ZigBee-specific layers with the same level of detail. The examples given throughout this book are generally referred to as ZigBee wireless networking examples; however, most of the discussions are still applicable even if only IEEE 802.15.4 PHY and MAC layers are implemented.

1.5   Frequencies of Operation and Data Rates

There are three frequency bands in the latest version of IEEE 802.15.4, which was released in September 2006:

- 868–868.6 MHz (868 MHz band)

- 902–928 MHz (915 MHz band)

- 2400–2483.5 MHz (2.4 GHz band)

The 868 MHz band is used in Europe for a number of applications, including short-range wireless networking [11]. The other two bands (915 MHz and 2.4 GHz) are part of industrial,

Table 1.1: IEEE 802.15.4 Data Rates and Frequencies of Operation

	Frequency (MHz)	Number of Channels	Modulation	Chip Rate (Kchip/s)	Bit Rate (Kb/s)	Symbol Rate (Ksymbol/s)	Spreading Method
	868–868.6	1	BPSK	300	20	20	Binary DSSS
	902–928	10	BPSK	600	40	40	Binary DSSS
Optional	868–868.6	1	ASK	400	250	12.5	20-bit PSSS
	902–928	10	ASK	1600	250	50	5-bit PSSS
Optional	868–868.6	1	O-QPSK	400	100	25	16-array orthogonal
	902–928	10	O-QPSK	1000	250	62.5	16-array orthogonal
	2400–2483.5	16	O-QPSK	2000	250	62.5	16-array orthogonal

scientific, and medical (ISM) frequency bands. The 915 MHz frequency band is used mainly in North America, whereas the 2.4 GHz band is used worldwide.

Table 1.1 provides further details regarding the ways these three frequency bands are used in the IEEE 802.15.4 standard. IEEE 802.15.4 requires that if a transceiver supports the 868 MHz band, it must support 915 MHz band as well, and vice versa. Therefore, these two bands are always bundled together as the 868/915 MHz frequency bands of operation.

IEEE 802.15.4 has one mandatory and two optional specifications for the 868/915 MHz bands. The mandatory requirements are simpler to implement but yield lower data rates (20 Kbps and 40 Kbps, respectively). Before the introduction of two optional PHY modes of operation in 2006, the only way to have a data rate better than 40 Kbps was to utilize the 2.4 GHz frequency band. With the addition of two new PHYs, if for any reason (such as existence of strong interference in the 2.4 GHz band) it is not possible to operate in the 2.4 GHz band, or if the 40 Kbps data rate is not sufficient, the user now has the option to achieve the 250 Kbps data rate at the 868/915 MHz bands.

If a user chooses to implement the optional modes of operation, IEEE 802.15.4 still requires that it accommodate the low-data-rate mandatory mode of operation in the

868/915 MHz bands as well. Also, the transceiver must be able to switch dynamically between the mandatory and optional modes of operation in 868/915 MHz bands.

A 2.4 GHz transceiver may support 868/915 MHz bands, but it is not required by IEEE 802.15.4. There is room for only a single channel in the 868 MHz band. The 915 MHz band has 10 channels (excluding the optional channels). The total number of channels in the 2.4 GHz band is 16.

The 2.4 GHz ISM band is accepted worldwide and has the maximum data rate and number of channels. For these reasons, developing transceivers for the 2.4 GHz band is a popular choice for many manufacturers. However, IEEE 802.11b operates in the same band and the coexistence can be an issue in some applications. (The coexistence challenge is covered in Chapter 8.) Also, the lower the frequency band is, the better the signals penetrate walls and various objects. Therefore, some users may find the 868/915 MHz band a better choice for their applications.

There are three modulation types in IEEE 802.15.4: binary phase shift keying (BPSK), amplitude shift keying (ASK), and offset quadrature phase shift keying (O-QPSK). In BPSK and O-QPSK, the digital data is in the phase of the signal. In ASK, in contrast, the digital data is in the amplitude of the signal.

All wireless communication methods in IEEE 802.15.4 (Table 1.1) take advantage of either direct sequence spread spectrum (DSSS) or parallel sequence spread spectrum (PSSS) techniques. DSSS and PSSS help improve performance of receivers in a multipath environment [12].

The basics of DSSS and PSSS spreading methods, as well as different modulations techniques and symbol-to-chip mappings, are covered in Chapter 4. The multipath issue and radio frequency (RF) propagation characteristics are covered in Chapter 5.

1.6   Interoperability

ZigBee has a wide range of applications; therefore, several manufacturers provide ZigBee-enabled solutions. It is important for these ZigBee-based devices be able to interact with each other regardless of the manufacturing origin. In other words, the devices should be *interoperable*. Interoperability is one of the key advantages of the ZigBee protocol stack. ZigBee-based devices are interoperable even when the messages are encrypted for security reasons.

1.7   Device Types

There are two types of devices in an IEEE 802.15.4 wireless network: *full-function devices* (FFDs) and *reduced-function devices* (RFDs). An FFD is capable of performing all the duties described in the IEEE 802.15.4 standard and can accept any role in the network. An RFD, on the other hand, has limited capabilities. For example, an FFD can communicate with any other device in a network, but an RFD can talk only with an FFD device. RFD devices are intended for very simple applications such as turning on or off a switch. The processing power and memory size of RFD devices are normally less than those of FFD devices.

1.8   Device Roles

In an IEEE 802.15.4 network, an FFD device can take three different roles: coordinator, PAN coordinator, and device. A *coordinator* is an FFD device that is capable of relaying messages. If the coordinator is also the principal controller of a personal area network (PAN), it is called a *PAN coordinator*. If a device is not acting as a coordinator, it is simply called a *device*.

The ZigBee standard uses slightly different terminology (see Figure 1.5). A ZigBee *coordinator* is an IEEE 802.15.4 PAN coordinator. A ZigBee *router* is a device that can act as an IEEE 802.15.4 coordinator. Finally, a ZigBee *end device* is a device that is neither a coordinator nor a router. A ZigBee end device has the least memory size and fewest processing capabilities and features. An end device is normally the least expensive device in the network.

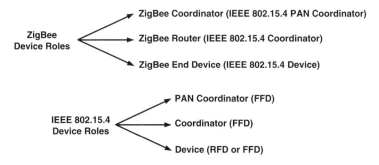

Figure 1.5: Device Roles in the IEEE 802.15.4 and ZigBee Standards

1.9   ZigBee Networking Topologies

The network formation is managed by the ZigBee networking layer. The network must be in one of two networking topologies specified in IEEE 802.15.4: star and peer-to-peer.

In the *star topology*, shown in Figure 1.6, every device in the network can communicate only with the PAN coordinator. A typical scenario in a star network formation is that an FFD, programmed to be a PAN coordinator, is activated and starts establishing its network. The first thing this PAN coordinator does is select a unique PAN identifier that is not used by any other network in its *radio sphere of influence*—the region around the device in which its radio can successfully communicate with other radios. In other words, it ensures that the PAN identifier is not used by any other nearby network.

In a *peer-to-peer topology* (see Figure 1.7), each device can communicate directly with any other device if the devices are placed close enough together to establish a successful communication link. Any FFD in a peer-to-peer network can play the role of the PAN coordinator. One way to decide which device will be the PAN coordinator is to pick the first FFD device that starts communicating as the PAN coordinator. In a peer-to-peer network, all the devices that participate in relaying the messages are FFDs because RFDs are not capable of relaying the messages. However, an RFD can be part of the network and communicate only with one particular device (a coordinator or a router) in the network.

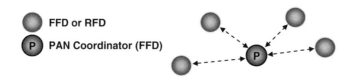

Figure 1.6: A Star Network Topology

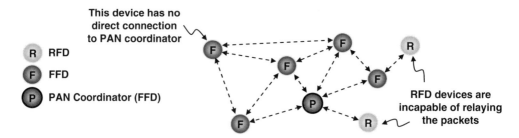

Figure 1.7: A Mesh Networking Topology

A peer-to-peer network can take different shapes by defining restrictions on the devices that can communicate with each other. If there is no restriction, the peer-to-peer network is known as a *mesh topology*. Another form of peer-to-peer network ZigBee supports is a *tree topology* (see Figure 1.8). In this case, a ZigBee coordinator (PAN coordinator) establishes the initial network. ZigBee routers form the branches and relay the messages. ZigBee end devices act as leaves of the tree and do not participate in message routing. ZigBee routers can grow the network beyond the initial network established by the ZigBee coordinator.

Figure 1.8 also shows an example of how relaying a message can help extend the range of the network and even go around barriers. For example, device A needs to send a message to device B, but there is a barrier between them that is hard for the signal to penetrate. The tree topology helps by relaying the message around the barrier and reach device B. This is sometimes referred to as *multihopping* because a message hops from one node to another until it reaches its destination. This higher coverage comes at the expense of potential high message latency.

An IEEE 802.15.4 network, regardless of its topology, is always created by a PAN coordinator. The PAN coordinator controls the network and performs the following minimum duties:

- Allocate a unique address (16-bit or 64-bit) to each device in the network.

- Initiate, terminate, and route the messages throughout the network.

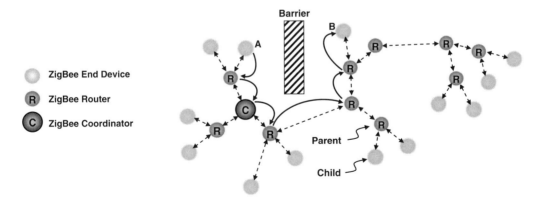

Figure 1.8: A ZigBee Tree Topology

- Select a unique PAN identifier for the network. This PAN identifier allows the devices within a network to use the 16-bit short-addressing method and still be able to communicate with other devices across independent networks.

There is only one PAN coordinator in the entire network. A PAN coordinator may need to have long active periods; therefore, it is usually connected to a main supply rather than a battery. All other devices are normally battery powered. The smallest possible network includes two devices: a PAN coordinator and a device.

1.10   ZigBee and IEEE 802.15.4 Communication Basics

This section reviews some communication basics such as multiple access method, data transfer methods, and addressing in IEEE 802.15.4 and ZigBee.

1.10.1   CSMA-CA

IEEE 802.15.4 implements a simple method to allow multiple devices to use the same frequency channel for their communication medium. The channel access mechanism used is Carrier Sense Multiple Access with Collision Avoidance (CSMA-CA). In CSMA-CA, anytime a device wants to transmit, it first performs a clear channel assessment (CCA) to ensure that the channel is not in use by any other device. Then the device starts transmitting its own signal. The decision to declare a channel clear or not can be based on measuring the spectral energy in the frequency channel of interest or detecting the type of the occupying signal.

When a device plans to transmit a signal, it first goes into receive mode to detect and estimate the signal energy level in the desired channel. This task is known an *energy detection* (ED). In ED, the receiver does not try to decode the signal, and only the signal energy level is estimated. If there is a signal already in the band of interest, ED does not determine whether or not this is an IEEE 802.15.4 signal.

An alternative way to declare a frequency channel clear or busy is *carrier sense* (CS). In CS, in contrast with ED, the type of the occupying signal is determined and, if this signal is an IEEE 802.15.4 signal, then the device may decide to consider the channel busy even if the signal energy is below a user-defined threshold.

If the channel is not clear, the device backs off for a random period of time and tries again. The random back-off and retry are repeated until either the channel becomes clear or the device reaches its user-defined maximum number of retries.

1.10.2   Beacon-Enabled vs. Nonbeacon Networking

There are two methods for channel access: contention based or contention free. In *contention-based channel access*, all the devices that want to transmit in the same frequency channel use the CSMA-CA mechanism, and the first one that finds the channel clear starts transmitting. In the *contention-free* method, the PAN coordinator dedicates a specific time slot to a particular device. This is called a *guaranteed time slot* (GTS). Therefore, a device with an allocated GTS will start transmitting during that GTS without using the CSMA-CA mechanism.

To provide a GTS, the PAN coordinator needs to ensure that all the devices in the network are synchronized. Beacon is a message with specific format that is used to synchronize the clocks of the nodes in the network. The format of the beacon frame is discussed in section 1.14.2.1.1. A coordinator has the option to transmit beacon signals to synchronize the devices attached to it. This is called a *beacon-enabled PAN*. The disadvantage of using beacons is that all the devices in the network must wake up on a regular basis, listen for the beacon, synchronize their clocks, and go back to sleep. This means that many of the devices in the network may wake up only for synchronization and not perform any other task while they are active. Therefore, the battery life of a device in a beacon-enabled network is normally less than a network with no beaconing.

A network in which the PAN coordinator does not transmit beacons is known as a *nonbeacon network*. A nonbeacon network cannot have GTSs and therefore contention-free periods because the devices cannot be synchronized with one another. The battery life in a nonbeacon network can be noticeably better than in a beacon-enabled network because in a nonbeacon network, the devices wake up less often.

1.10.3   Data Transfer Methods

There are three types of data transfer in IEEE 802.15.4:

- Data transfer to a coordinator from a device

- Data transfer from a coordinator to a device

- Data transfer between two peer devices

All three methods can be used in a peer-to-peer topology. In a star topology, only the first two are used, because no direct peer-to-peer communication is allowed.

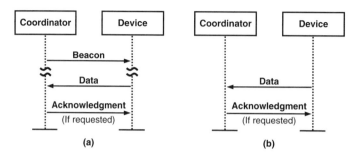

Figure 1.9: Data Transfer to a Coordinator in IEEE 802.15.4: (a) Beacon Enabled, and (b) Nonbeacon Enabled

1.10.3.1   Data Transfer to a Coordinator

In a beacon-enabled network, when a device decides to transmit data to the coordinator, the device synchronizes its clock on a regular basis and transmits the data to the coordinator using the CSMA-CA method (assuming that the transmission does not occur during a GTS). The coordinator may acknowledge the reception of the date only if it is requested by the data transmitter. This sequence chart is shown in Figure 1.9a.

Figure 1.9b shows the data transfer sequence in a nonbeacon-enabled network. In this scenario, the device transmits the data as soon as the channel is clear. The transmission of an acknowledgment by the PAN coordinator is optional.

1.10.3.2   Data Transfer from a Coordinator

Figure 1.10a illustrates the data transmission steps to transfer data from a coordinator to a device in a beacon-enabled network. If the coordinator needs to transmit data to a particular device, it indicates in its beacon message that a data message is pending for that device. The device then sends a data request message to the coordinator indicating that it is active and ready to receive the data. The coordinator acknowledges the receipt of the data request and sends the data to the device. Sending the acknowledgment by the device is optional.

In a nonbeacon-enabled network (Figure 1.10b), the coordinator needs to wait for the device to request the data. If the device requests the data but there is no data pending for that device, the coordinator sends an acknowledgment message with a specific format that indicates there is no data pending for that device. Alternatively, the coordinator may send a data message with a zero-length payload.

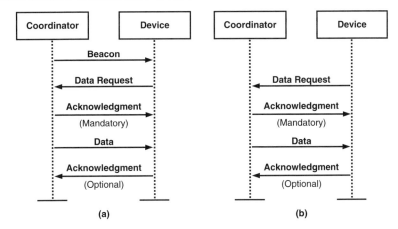

Figure 1.10: Data Transfer from a Coordinator to a Device: (a) Beacon Enabled, and (b) Nonbeacon Enabled

1.10.3.3 Peer-to-Peer Data Transfer

In a peer-to-peer topology, each device can communicate directly with any other device. In many applications, the devices engaged in peer-to-peer data transmissions and receptions are synchronized. (Further details regarding peer-to-peer communication are provided in Chapter 3.)

1.10.4 Data Verification

A *packet* is a number of bits transmitted together with a specific format. The receiver needs to have a mechanism to verify whether any of the received bits are recovered in error. IEEE 802.15.4 uses a 16-bit Frame Check Sequence (FCS) based on the International Telecommunication Union (ITU) Cyclic Redundancy Check (CRC) to detect possible errors in the data packet [13]. The details of CRC implementation are provided in Section 3.3.5.1.1.

1.10.5 Addressing

Each device in a network needs a unique address. IEEE 802.15.4 uses two methods of addressing:

- 16-bit short addressing
- 64-bit extended addressing

A network can choose to use either 16-bit or 64-bit addressing. The short address allows communication within a single network. Using the short addressing mechanism allows for a reduction in the length of the messages and saves on required memory space that is allocated for storing the addresses. The combination of a unique PAN identifier and a short address can be used for communication between independent networks.

Availability of 64-bit addressing means that the maximum number of devices in a network can be 2^{64}, or approximately 1.8×10^{19}. Therefore, an IEEE 802.15.4 wireless network has practically no limit on the number of devices that can join the network.

The Network (NWK) layer of the ZigBee protocol assigns a 16-bit NWK address in addition to the IEEE address. A simple lookup table is used to map each 64-bit IEEE address to a unique NWK address. The NWK layer transactions require the use of the NWK address.

Each radio in a network can have a single IEEE address and a single NWK address. But there can be up to 240 devices connected to a single radio. Each one of these devices is distinguished by a number between 1 and 240 known as the *endpoint* address.

1.11   Association and Disassociation

Association and *disassociation* are services provided by IEEE 802.15.4 that can be used to allow devices to join or leave a network. For example, when a device wants to join a PAN, it sends an association request to the coordinator. The coordinator can accept or reject the association request. The device uses the disassociation to notify the coordinator of its intent to leave the network.

1.12   Binding

Binding is the task of creating logical links between the applications that are related. For example, a ZigBee device connected to a lamp is logically related to another ZigBee device connected to the switch that controls the lamp. The information regarding these logical links is stored in a *binding table*. The ZigBee standard, at the application layer, provides support for creating and maintaining binding tables. Devices logically related in a binding table are called *bound devices*.

1.13   ZigBee Self-Forming and Self-Healing Characteristics

As discussed in Section 1.9, a ZigBee network starts its formation as soon as devices become active. In a mesh network, for example, the first FFD device that starts communicating can establish itself as the ZigBee coordinator, and other devices then join the network by sending association requests. Because no additional supervision is required to establish a network, ZigBee networks are considered *self-forming networks*.

On the other hand, when a mesh network is established, there is normally more than one way to relay a message from one device to another. Naturally, the most optimized way is selected to route the message. However, if one of the routers stops functioning due to exhaustion of its battery or if an obstacle blocks the message route, the network can select an alternative route. This is an example of the *self-healing* characteristic of ZigBee mesh networking.

ZigBee is considered an ad hoc wireless network. In an ad hoc wireless network, some of the wireless nodes are willing to forward data for other devices. The route that will carry a message from the source to the destination is selected dynamically based on the network connectivity. If the network condition changes, it might be necessary to change the routing in the network. This is in contrast to some other networking technologies in which there is an infrastructure in place, and some designated devices always act as routers in the network.

1.14 ZigBee and IEEE 802.15.4 Networking Layer Functions

This section provides a functional overview of the ZigBee and IEEE 802.15.4 protocol layers. The details are provided in Chapter 3.

1.14.1 PHY Layer

In ZigBee wireless networking (Figure 1.3), the lowest protocol layer is the IEEE 802.15.4 Physical layer, or PHY. This layer is the closest layer to hardware and directly controls and communicates with the radio transceiver. The PHY layer is responsible for activating the radio that transmits or receives packets. The PHY also selects the channel frequency and makes sure the channel is not currently used by any other devices on another network.

1.14.1.1 PHY Packet General Structure

Data and commands are communicated between various devices in the form of packets. The general structure of a packet is shown in Figure 1.11. The PHY packet consists of three components: the Synchronization header (SHR), the PHY header (PHR), and the PHY payload.

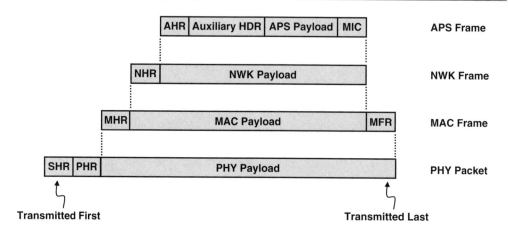

Figure 1.11: ZigBee Packet Structure

The SHR enables the receiver to synchronize and lock into the bit stream. The PHR contains frame length information, and the PHY payload is provided by upper layers and includes data or commands for the recipient device.

The MAC frame, which is transmitted to other devices as a PHY payload, has three sections. The MAC header (MHR) contains information such as addressing and security. The MAC payload has a variable length size (including zero length) and contains commands or data. The MAC footer (MFR) contains a 16-bit Frame Check Sequence (FCS) for data verification.

The NWK frame has two parts: the NWK header (NHR) and the NWK payload. The NWK header has network-level addressing and control information. The NWK payload is provided by the APS sublayer. In the APS sublayer frame, the APS header (AHR) has application-layer control and addressing information. The auxiliary frame header (auxiliary HDR) contains the mechanism used to add security to the frame and the security keys used. These security keys are shared among the corresponding devices and help unlock the information. The NWK and MAC frames can also have optional auxiliary headers for additional security. The APS payload contains data or commands. The Message Integrity Code (MIC) is a security feature in the APS frame that is used to detect any unauthorized change in the content of the message.

Figure 1.11 shows that the first transmitted bit is the *least significant bit* (LSB) of the SHR. The *most significant bit* (MSB) of the last octet of the PHY payload is transmitted last.

1.14.2 MAC Layer

The Medium Access Control (MAC) layer provides the interface between the PHY layer and the NWK layer. The MAC is responsible for generating beacons and synchronizing the device to the beacons (in a beacon-enabled network). The MAC layer also provides association and disassociation services.

1.14.2.1 MAC Frame Structures

The IEEE 802.15.4 defines four MAC frame structures:

- Beacon frame

- Data frame

- Acknowledge frame

- MAC command frame

The beacon frame is used by a coordinator to transmit beacons. The beacons are used to synchronize the clock of all the devices within the same network. The data and acknowledgment frames are used to transmit data and accordingly acknowledge the successful reception of a frame. The MAC commands are transmitted using a MAC command frame.

1.14.2.1.1 The Beacon Frame The structure of a beacon frame is shown in Figure 1.12. The entire MAC frame is used as a payload in a PHY packet. The content of the PHY payload is referred to as the PHY Service Data Unit (PSDU).

In the PHY packet, the preamble field is used by the receiver for synchronization. The start-of-frame delimiter (SDF) indicates the end of SHR and start of PHR. The frame length specifies the total number of octets in the PHY payload (PSDU).

The MAC frame consists of three sections: the MAC header (MHR), the MAC payload, and the MAC footer (MFR). The frame control field in the MHR contains information defining the frame type, addressing fields, and other control flags. The sequence number specifies the beacon sequence number (BSN). The addressing field provides the source and destination addresses. The auxiliary security header is optional and contains information required for security processing.

The MAC payload is provided by the NWK layer. The *superframe* is a frame bounded by two beacon frames. The superframe is optionally used in a beacon-enabled network and

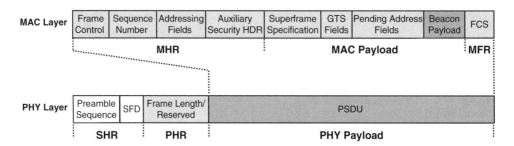

Figure 1.12: The MAC Beacon Frame Structure

helps define GTSs. The GTS field in the MAC payload determines whether a GTS is used to receive or transmit.

The beacon frame is not only used to synchronize the devices in a network but is also used by the coordinator to let a specific device in a network know there is data pending for that device in the coordinator. The device, at its discretion, will contact the coordinator and request that it transmit the data to the device. This is called *indirect transmission.* The pending address field in the MAC payload contains the address of the devices that have data pending in the coordinator. Every time a device receives a beacon, it will check the pending address field to see if there is data pending for it.

The beacon payload field is an optional field that can be used by the NWK layer and is transmitted along with the beacon frame. The receiver uses the Frame Check Sequence (FCS) field to check for any possible error in the received frame. Further details of the frame formats are provided in Chapter 3.

1.14.2.1.2 The Data Frame The MAC data frame is shown in Figure 1.13. The data payload is provided by the NWK layer. The data in the MAC payload is referred to as the MAC Service Data Unit (MSDU). The fields in this frame are similar to the beacon frame except the superframe, GTS, and pending address fields are not present in the MAC data frame. The MAC data frame is referred to as the MAC Protocol Data Unit (MPDU) and becomes the PHY payload.

1.14.2.1.3 The Acknowledgment Frame The MAC acknowledgment frame, shown in Figure 1.14, is the simplest MAC frame format and does not carry any MAC payload. The acknowledgment frame is sent by one device to another to confirm successful reception of a packet.

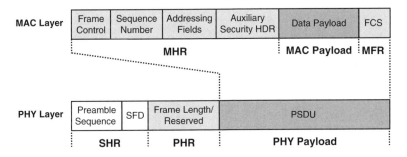

Figure 1.13: The MAC Data Frame Structure

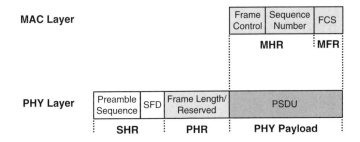

Figure 1.14: The MAC Acknowledgment Frame Structure

1.14.2.1.4 The Command Frame The MAC commands such as requesting association or disassociation with a network are transmitted using the MAC command frame (see Figure 1.15). The command type field determines the type of the command (e.g., association request or data request). The command payload contains the command itself. The entire MAC command frame is placed in the PHY payload as a PSDU.

1.14.3 The NWK Layer

The NWK layer interfaces between the MAC and the APL and is responsible for managing the network formation and routing. *Routing* is the process of selecting the path through which the message will be relayed to its destination device. The ZigBee coordinator and the routers are responsible for discovering and maintaining the routes in the network. A ZigBee end device cannot perform route discovery. The ZigBee coordinator or a router will perform route discovery on behalf of the end device. The NWK layer of a ZigBee coordinator is responsible for establishing a new network and selecting the network topology (tree, star, or mesh). The ZigBee coordinator also assigns the NWK addresses to the devices in its network.

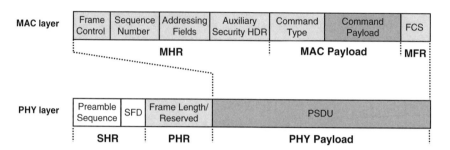

Figure 1.15: The MAC Command Frame Structure

1.14.4   The APL Layer

The application (APL) layer is the highest protocol layer in the ZigBee wireless network and hosts the application objects. Manufacturers develop the application objects to customize a device for various applications. Application objects control and manage the protocol layers in a ZigBee device. There can be up to 240 application objects in a single device.

The ZigBee standard offers the option to use application profiles in developing an application. An *application profile* is a set of agreements on application-specific message formats and processing actions. The use of an application profile allows further interoperability between the products developed by different vendors for a specific application. If two vendors use the same application profile to develop their products, the product from one vendor will be able to interact with products manufactured by the other vendor as though both were manufactured by the same vendor.

1.14.5   Security

In a wireless network, the transmitted messages can be received by any nearby device, including an intruder. There are two main security concerns in a wireless network. The first one is *data confidentiality*. The intruder device can gain sensitive information by simply listening to the transmitted messages. Encrypting the messages before transmission will solve the confidentiality problem. An encryption algorithm modifies a message using a string of bits known as the *security key,* and only the intended recipient will be able to recover the original message. The IEEE 802.15.4 standard supports the use of Advanced Encryption Standard (AES) [14] to encrypt their outgoing messages.

The second concern is that the intruder device may modify and resend one of the previous messages even if the messages are encrypted. Including a message integrity code (MIC) with each outgoing frame will allow the recipient to know whether the message has been changed in transit. This process is known as *data authentication*.

One of the main constraints in implementing security features in a ZigBee wireless network is limited resources. The nodes are mainly battery powered and have limited computational power and memory size. ZigBee is targeted for low-cost applications and the hardware in the nodes might not be tamper resistant. If an intruder acquires a node from an operating network that has no tamper resistance, the actual key could be obtained simply from the device memory. A tamper-resistant node can erase the sensitive information, including the security keys, if tampering is detected.

1.15 The ZigBee Gateway

A ZigBee gateway provides the interface between a ZigBee network and another network using a different standard. For example, if ZigBee wireless networking is used to gather patient information locally inside a room, the information might need to be transmitted over the Internet to a monitoring station. In this case, the ZigBee gateway implements both the ZigBee protocol and the Internet protocol to be able to translate ZigBee packets to Internet protocol packet format, and vice versa.

1.16 ZigBee Metaphor

One of the key characteristics of the ZigBee standard is its mesh networking capability. In a large distributed mesh network, a message is relayed from one device to another until it reaches its faraway destination. Similarly, when a group of honey bees, distributed in a large field, want to communicate a message all the way back to their hive, they use message relaying. Each bee performs a specific zigzag dance, which is repeated by the next bee that is slightly closer to the hive. This process is repeated until the message gets to the hive. The name *ZigBee* was selected as a metaphor for the way devices on the network find and interact with one another [15].

References

[1] ZigBee Specification 053474r17, Jan. 2008; available from www.zigbee.org.
[2] S. Dagtas et al., "Multi-stage Real Time Health Monitoring via ZigBee in Smart Homes," *Proceedings of 2007 IEEE International Conference on Advanced Information Networking and Applications Workshops* (AINAW), pp. 782–786.

[3] J. P. Lynch, "An overview of wireless structural health monitoring for civil structures," *Phil. Trans. R. Soc. A*, 2007, pp. 345–372.

[4] ZigBee Alliance, available at www.zigbee.org.

[5] IEEE 802.15.4: Wireless Medium Access Control (MAC) and Physical Layer (PHY) Specifications for Low-Rate Wireless Personal Area Networks (WPANs), Sept. 2006.

[6] IEEE 802.11: Wireless LAN Medium Access Control (MAC) and Physical Layer (PHY) Specifications.

[7] J. Gutierrez, et al., "*Low-Rate Wireless Personal Area Networks*," IEEE Press, 2007.

[8] IEEE 802.15.3-2003: Wireless Medium Access Control (MAC) and Physical Layer (PHY) Specifications for High-Rate Wireless Personal Area Networks (WPANs).

[9] Open Systems Interconnection Basic Reference Model: The Basic Model, ISO/IEC 7498-1:1994.

[10] IEEE 802 LAN/MAN standards committee, available at www.ieee802.org.

[11] European Radiocommunications Committee (ERC)/CEPT Report 98, available at www.learnZigBee.com.

[12] H. Schwetlick, et al., "PSSS—Parallel Sequence Spread Spectrum: A Physical Layer for RF Communication," *IEEE International Symposium on Consumer Electronics*, 2004, pp. 262–265.

[13] International Telecommunication Union, available at www.itu.int.

[14] Advanced Encryption Standard (AES), Federal Information Processing Standards Publication 197, U.S. Department of Commerce/N.I.S.T, Springfield, Virginia, Nov. 26, 2001. Available at http://csrc.nist.gov/.

[15] J. Adams, et al., "Busy as a Bee," *IEEE Spectrum,* Oct. 2006, available at http://spectrum.ieee.org.

CHAPTER 2

ZigBee/IEEE 802.15.4 Networking Examples

ZigBee networking has a diverse range of applications, including but not limited to home automation, inventory tracking, and healthcare. This chapter reviews a number of the application scenarios in which ZigBee devices can increase efficiency and/or reduce cost. Full ZigBee protocol implementation has the advantage of reliable mesh networking capability. However, if the application is simple, it might be possible to implement only IEEE 802.15.4 layers.

2.1 Home Automation

Home automation is one of the major application areas for ZigBee wireless networking. In this section, a number of these use cases are reviewed. The typical data rate in home automation is only 10 Kbps [1]. Figure 2.1 shows some of the possible ZigBee applications in a typical residential building. Most of the applications shown in Figure 2.1 are briefly reviewed in this chapter.

2.1.1 Security Systems

A security system can consist of several sensors, including motion detectors, glass-break sensors, and security cameras. These devices need to communicate with the central security panel through either wire or a wireless network. ZigBee-based security systems simplify installing and upgrading security systems [2]. Despite ZigBee's low data rate, it is still possible to transfer images wirelessly with acceptable quality. For example, ZigBee has been used in a wireless camera system that records videos of visitors at a home's front door and transmits them to a dedicated monitor inside the house.

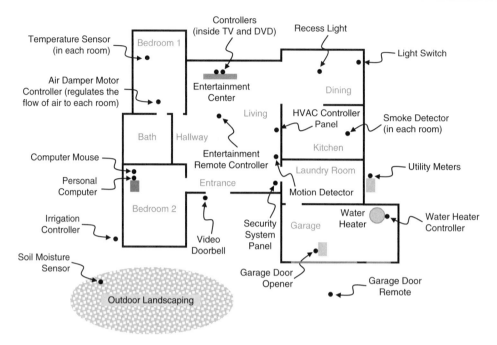

Figure 2.1: Possible ZigBee-Enabled Devices in a Typical Residential Building

2.1.2 Meter-Reading Systems

Utility meters need to be read on a regular basis to generate utility bills. One way to do so is to read the meters manually at homeowners' premises and enter the values into a database. A ZigBee-based automatic meter-reading (AMR) system can create self-forming wireless mesh networks across residential complexes that link meters with utilities' corporate offices. AMR provides the opportunity to remotely monitor a residence's electric, gas, and water usage and eliminate the need for a human visiting each residential unit on a monthly basis.

An AMR can do more than simply deliver the total monthly usage data; it can gather detailed usage information, automatically detect leaks and equipment problems, and help in tamper detection [3]. ZigBee-based wireless devices not only perform monitoring tasks, they can manage the usage peak by communicating with the appliances inside the house. For example, when there is a surge in electricity usage, a ZigBee-enabled electric water heater can be turned off for a short period of time to reduce the peak power consumption.

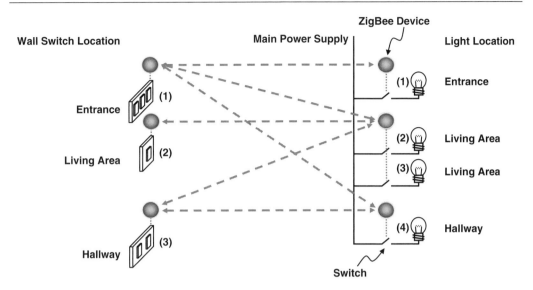

Figure 2.2: Light Control in a Residential Building using ZigBee Wireless Networking

2.1.3 Irrigation Systems

A sensor-based irrigation system can result in efficient water management. Sensors across the landscaping field can communicate to the irrigation panel the soil moisture level at different depths. The controller determines the watering time based on moisture level, plant type, time of day, and the season. A distributed wireless sensor network eliminates the difficulty of wiring sensor stations across the field and reduces the maintenance cost.

2.1.4 Light Control Systems

Light control is one of the classic examples of using ZigBee in a house or commercial building. In traditional light installation, to turn on or off the light it is necessary to bring the wire from the light to a switch. Installation of a new recess light, for example, requires new wiring to a switch. If the recess light and the switch are equipped with ZigBee devices, no wired connection between the light and the switch is necessary. In this way, any switch in the house can be assigned to turn on and off a specific light.

Figure 2.2 is an example of wireless connections between wall switches and lights. In our example, the lights are located in a residential building entrance, living area, and hallway. The wall switch in the entrance can turn on and off any of the four lights. The living area wall switch, in contrast, communicates only with the lights in the living area. Living area lights are in close proximity to each other, and therefore a single ZigBee device can be used for both lights.

The concept of using binding tables (see Section 1.12) is applicable in the example of Figure 2.2. Wall switch 1 is logically connected to all four lights. Wall switch 2 is bound only to the lights in the living area. One of the devices in the network has the task of storing and updating the binding table.

A ZigBee-enabled recess light can be more expensive than a regular recess light, but the installation cost of a ZigBee-enabled light is lower because it requires no additional wiring to a wall switch. Using wireless remotes to control the lights is not a new concept. ZigBee provides the opportunity to implement this concept on a large scale by ensuring long battery life and interoperability of products from various vendors in a reliable and low-cost network.

In addition to potential cost savings, ZigBee-enabled lights can have other benefits in a house. For example, the ZigBee devices embedded in the recess lights can act as routers to relay a message across the house, or the lights can be programmed to dim whenever the television set is turned on. The ZigBee light control mechanism has been used for street light controls as well [4].

2.1.5   Multizone HVAC Systems

The multizone control system allows a single heating, ventilation, and air-conditioning (HVAC) unit to have separate temperature zones in the house. Zoning the HVAC system can help save energy by controlling the flow of air to each room and avoiding cooling or heating unnecessary areas. Figure 2.3 is a simplified diagram that shows motors controlling

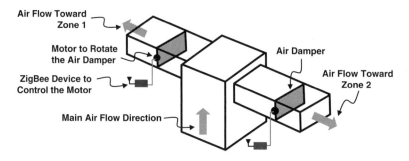

Figure 2.3: Multizone Air Conditioning using ZigBee-Controlled Air Dampers

air dampers and regulating the flow of air to different zones. ZigBee devices control these motors based on the commands they receive from the main HVAC zone control panel and temperature sensors. An alternative way of implementing a multizone control system is to connect the zone control panel, motors, and temperature sensors via wires instead of a wireless network. A wired system has less flexibility and additional labor cost for wiring, but the cost of the parts might be slightly lower. Total system cost and flexibility for future modifications should be the decision factors in selecting between these two implementation methods.

2.2 Consumer Electronics: Remote Control

In consumer electronics, ZigBee can be used in wireless remote controls, game controllers, a wireless mouse for a personal computer, and many other applications. In this section, we briefly review the application of ZigBee in wireless remotes.

An infrared (IR) remote controller communicates with televisions, DVDs, and other entertainment devices via infrared signals. The limitation of IR remotes is that they provide only one-way communication from the remote to the entertainment device. Also, IR signals do not penetrate walls and other objects and therefore require line of sight to operate properly. Radio frequency (RF) signals, however, easily penetrate walls and most objects.

IEEE 802.15.4 is a proper replacement for IR technology in remote controls because of the low cost and long battery life of ZigBee-based wireless communication [5]. IEEE 802.15.4 can be used to create two-way communications between the remote control and the entertainment device. For example, song information or on-screen programming options can be downloaded in to the remote itself, even when the remote control is not in the same room as the entertainment device.

2.3 Industrial Automation

At the industrial level, ZigBee mesh networking can help in areas such as energy management, light control, process control, and asset management. In this section, application of ZigBee in asset management and personnel tracking is briefly reviewed.

2.3.1 Asset Management and Personnel Tracking

Passive radio frequency identification (RFID) tags have been in use for several years. Although a passive RFID tag does not have any battery, the RFID reader unit can be a battery-powered instrument. A passive RFID tag can transmit only simple information such as an ID number, which is sufficient for many asset management applications.

Active RFIDs, such as ZigBee devices, are battery powered and generally are more expensive than passive RFIDs. ZigBee-based active RFIDs have longer range than passive RFIDs and can provide additional services such as estimating the location of assets or personnel. Chapter 7 covers the details of ZigBee-based location methods. The basic concept of location estimation is shown in Figure 2.4, where location of personnel is tracked inside a typical office building with offices and cubes. There are three fixed ZigBee nodes with known locations. The mobile ZigBee node, carried by an employee, broadcasts a signal that is received by all three fixed nodes. The signal becomes weaker as it travels longer; therefore, the amplitude of the signal received by each of the fixed nodes can be different. There are several algorithms that can take the received signal strength at the three fixed nodes and calculate the approximate location of the mobile node. The signal transmitted from the mobile node is reflected from walls and other objects in the room before it reaches the fixed nodes, which results in reduced accuracy of the location estimation. Chapter 7 reviews some of the methods developed to improve the location estimation accuracy.

2.3.2   Livestock Tracking

Livestock are vulnerable to disease, and it is important to track and identify a diseased animal quickly. Rapid disease response reduces the number of producers impacted by a

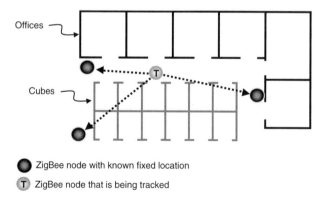

Offices

Cubes

● ZigBee node with known fixed location
Ⓣ ZigBee node that is being tracked

**Figure 2.4: Personnel Tracking in an Office Building using ZigBee
Wireless Networking**

disease outbreak or other animal health events [6]. Passive RFID tags have been used as an inexpensive solution for livestock tracking and can be sufficient for some applications. Passive RFID tags have limited range and can only provide previously stored information such as an identification number. IEEE 802.15.4-based active tags can cost more than passive ones, but the IEEE 802.15.4 tags have extended range and can provide additional information such as animal heartbeat and the animal's approximate location.

2.4   Healthcare

One of the applications of IEEE 802.15.4 in the healthcare industry is monitoring a patient's vital information remotely. Consider a patient who is staying at his home but for whom it is important that his physician monitor his heart rate and blood pressure continuously. In this system, an IEEE 802.15.4 network can be used to collect data from various sensors connected to the patient. The 802.15.4 standard uses 128-bit Advanced Encryption Standard (AES) technology to securely transfer data between ZigBee devices and other networks.

Figure 2.5 is a simplified diagram of a remote monitoring system. A patient wears a ZigBee device that interfaces with a sensor, such as a blood pressure sensor, that gathers the information on a periodic basis. Then this information is transmitted to a ZigBee gateway. A ZigBee gateway provides the interface between a ZigBee network and other networks, such as an Internet Protocol (IP) network. The patient information is then transmitted over the Internet to a personal computer that the physician or nurse uses to monitor the patient. This system could help hospitals improve patient care and relieve hospital overcrowding by enabling them to monitor patients at home.

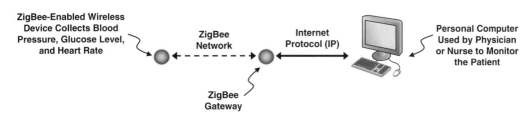

Figure 2.5: In-Home Patient Monitoring using ZigBee Wireless Networking

2.5   Other Applications

2.5.1   *Hotel Guest Room Access*

ZigBee-based systems can replace the magnetic key card systems used in hotels to access guest rooms. The traditional room access plastic cards have a magnetic strip on their back; the card reader installed on the guest door reads the information encoded into the magnetic strip to allow or deny access to the room. Installing this reader for each door requires wiring through the door. Alternatively, a ZigBee- based room access system includes a portable ZigBee device that acts as the key and a battery-powered ZigBee device inside the door that locks and unlocks it. Unlike the traditional method, the ZigBee-based room access system does not require wiring each door, which reduces the installation cost.

2.5.2   *Fire Extinguishers*

Fire extinguishers should be checked every 30 days to make sure all the canisters are charged and pressures are correct. Instead of checking the extinguishers manually, in a ZigBee-based monitoring system a sensor is attached to each extinguisher to monitor its status and wirelessly communicate with the coordinator when maintenance is needed. A ZigBee-based monitoring system not only saves time and labor cost, it also helps improve fire safety by immediately alerting authorities if a fire extinguisher is not working properly.

References

[1] J. Gutierrez et al., "*Low-Rate Wireless Personal Area Networks,*" IEEE Press, 2007.

[2] Home Heartbeat (EATON) Zigbee-Based Security System; available at www.home-heartbeat.com

[3] Automatic Meter Reading Association (AMRA); available at www.amra-intl.org.

[4] J. D. Lee et al., "Development of ZigBee-Based Street Light Control System," *Proceedings of IEEE Power System Symposium and Exposition*, pp. 2236–2240.

[5] Freescale Semiconductor, available at www.freescale.com/zigbee.

[6] National Animal Identification System (NAIS); available at www.usda.gov/nais.

CHAPTER 3

ZigBee and IEEE 802.15.4 Protocol Layers

Chapter 1 reviewed the basics of ZigBee and IEEE 802.15.4 wireless networking. This chapter provides further insights into the structure and services provided by each layer of the ZigBee and IEEE 802.15.4 standards. The protocol layers cooperate with each other to perform various tasks, such as joining a network or routing messages. The concept of service primitives, a convenient way of describing protocol services, is reviewed in this chapter. Although the chapter offers details on subjects such as frame formats, the emphasis is always on the functional descriptions of services provided by each protocol layer.

3.1   ZigBee and IEEE 802.15.4 Networking Layers

ZigBee wireless networking protocol layers are shown in Figure 3.1. The ZigBee protocol layers are based on the International Standards Organization (ISO) Open System Interconnect (OSI) basic reference model [1]. There are seven layers in the ISO/OSI model, but ZigBee implements only the layers that are essential for low-power, low-data-rate wireless networking. The lower two layers (PHY and MAC) are defined by the IEEE 802.15.4 standard [2]. The NWK and APL layers are defined by the ZigBee standard [3]. The security features are defined in both standards. A network that implements all of the layers in Figure 3.1 is considered a ZigBee wireless network.

Each layer communicates with the adjacent layers through service access points (SAPs). A SAP is a conceptual location at which one protocol layer can request the services of another protocol layer. For example, in Figure 3.1, the PHY Data Service Access Point (PD-SAP) is where the MAC layer requests any data service from the PHY layer.

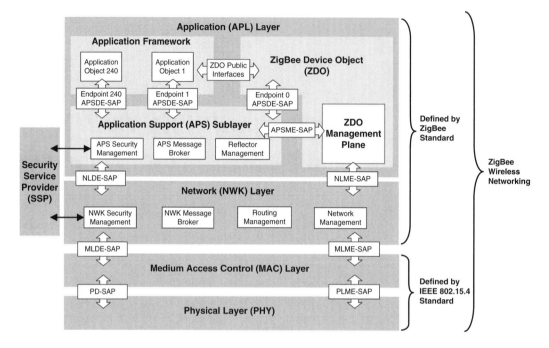

Figure 3.1: ZigBee Networking Protocol Layers

3.2 The IEEE 802.15.4 PHY Specifications

The IEEE 802.15.4 not only specifies the PHY protocol functions and interactions with the MAC layer, it also defines the minimum hardware-level requirements, such as the receiver sensitivity and the transmitter output power. The commercially available transceivers, however, can perform beyond the minimum requirements of the IEEE 802.15.4. Chapter 4 discusses the transceiver performance requirements and practical considerations. To avoid repeating the same material in two chapters, we cover all hardware-level requirements of the IEEE 802.15.4 PHY, including but not limited to the Power Spectral Density (PSD) mask, the Error Vector Magnitude (EVM), and the jamming resistance requirements, in Chapter 4.

3.2.1 Channel Assignments

The frequency channels are defined through a combination of channel numbers and channel pages. Channel page is a concept added to IEEE 802.15.4 in 2006 to distinguish

Table 3.1: Channel Assignments

Channel Page	Channel Number	Description
0	0	868 MHz band (BPSK)
	1–10	915 MHz band (BPSK)
	11–26	2.4 GHz band (O-QPSK)
1	0	868 MHz band (ASK)
	1–10	915 MHz band (ASK)
	11–26	Reserved
2	0	868 MHz band (O-QPSK)
	1–10	915 MHz band (O-QPSK)
	11–26	Reserved
3–31	Reserved	Reserved

between supported PHYs. In previous releases of IEEE 802.15.4 standard, the frequency channels were simply identified by channel numbers and there were no optional PHYs. In the initial release, there was no provision for more than a total of 27 channels, and hence PHYs implementing multiple operating frequency bands could not be supported. Each channel page can have a maximum of 27 channels. Table 3.1 shows the channel assignments in the IEEE 802.15.4 standard. The channel pages 0–2 are currently used for 868/915 MHz and 2.4 GHz bands. The channel pages 3–31 are reserved for future potential uses.

The channel page 0 supports all the channels defined in 2003 edition of IEEE 802.15.4. The channel pages 1 and 2 are used by the optional PHYs introduced in 2006 edition of the standard.

Each channel is identified by a channel number. In the first three channel pages, the channel number 0 is assigned to the 868 MHz band with the center frequency of 868.3 MHz. For other frequency bands, the lowest channel number is 1, which is assigned to the channel with the lowest frequency.

The center frequencies of the channels in the 915 MHz band can be calculated from the following equation:

$$\text{Center Frequency (MHz)} = 906 + 2 \times (\text{Channel Number} - 1)$$

Similarly, for the 2.4 GHz band, the center frequencies are calculated from the following equation:

$$\text{Center Frequency (MHz)} = 2405 + 5 \times (\text{Channel Number} - 11)$$

For example, the center frequencies of channel numbers 5 and 14 are 914 MHz and 2420 MHz, respectively.

3.2.2 Energy Detection

When a device plans to transmit a message, it first goes into the receive mode to detect and estimate the signal energy level in the desired channel. This task is known as *energy detection* (ED). The signal energy in the band of interest is averaged over eight symbol periods. In ED, the receiver does not attempt to detect the signal type; just the signal energy level is estimated. In other words, if a signal is occupying the frequency band of interest, performing an ED does not reveal whether this signal is an IEEE 802.15.4 standard-compliant signal or not.

The ED procedure might not be able to detect weak signals with energy levels close to the receiver sensitivity level. The receiver sensitivity is the lowest signal energy level that the receiver can successfully detect and demodulate with a packet error rate of less than 1%. The IEEE 802.15.4 allows 10 dB difference between the required receiver sensitivity level and the required energy detection level. Therefore, a receiver performing an ED must be able to detect and measure the energy of the signals as low as 10 dB above its required sensitivity level. For example, if the required receiver sensitivity is −85 dBm, the ED procedure must be able to detect and measure the energy of signals as low as −75 dBm. The ED range must be at least 40 dB, which for the same example translates to −75 dBm to −35 dBm.

The MAC requests the PHY to perform ED. The PHY returns an 8-bit integer indicating the energy level in the frequency channel of interest. The energy-level accuracy must be ±6 dB or better.

3.2.3 Carrier Sense

Similar to ED, *carrier sense* (CS) is a way of verifying whether a frequency channel is available to use. In CS, when a device plans to transmit a message, it first goes into the receive mode to detect the *type* of any possible signal that might be present in the desired frequency channel. In contrast with ED, in CS the signal is demodulated to verify whether the signal modulation and spreading are compliant with the characteristics of the PHY that is currently in use by the device. If the occupying signal is compliant to the IEEE 802.15.4 PHY, the device might choose to consider the channel busy regardless of the signal energy level.

3.2.4 Link Quality Indicator

The *link quality indicator* (LQI) is an indication of the quality of the data packets received by the receiver. The received signal strength (RSS) can be used as a measure of the signal quality. The RSS is a measure of the total energy of the received signal. The ratio of the desired signal energy to the total in-band noise energy (the *signal-to-noise ratio*, or SNR) is another way to judge the signal quality. As a general rule, higher SNR translates to lower chance of error in the packet. Therefore, a signal with high SNR is considered a high-quality signal. The link quality can also be judged using both the signal energy and the signal-to-noise ratio.

The LQI measurement is performed for each received packet. The LQI must have at least eight unique levels. The LQI is reported to the MAC layer and is available to the NWK and the APL layers for any type of analysis. For example, the NWK layer can use the reported LQI levels of the devices in the network to decide which path to use to route a message. In general, the path that has the highest overall LQI has a better chance of delivering a message to the destination. The LQI is only one of the decision factors in selecting a path to route a message. Other factors, such as routing energy efficiency considerations, can also influence the route selection. For example, a battery-powered device might be in an excellent location in terms of the link quality, but routing the messages frequently through this device will drain its battery much earlier than the rest of the devices in the same network.

3.2.5 Clear Channel Assessment

In the first step of the Carrier Sense Multiple Access with Collision Avoidance (CSMA-CA) channel access mechanism, the MAC requests the PHY to perform a clear channel assessment (CCA) to ensure that the channel is not in use by any other device. The CCA is part of the PHY management service (Section 3.2.7.2). In a CCA, the results of ED or CS can be used to decide whether a frequency channel should be considered available or busy. The CCA period must be eight symbols.

There are three CCA modes, and an IEEE 802.15.4-compliant PHY must be able to operate in any one of them:

- *CCA mode 1.* In this mode, only the ED result is taken into account. If the energy level is above the ED threshold, the channel is considered busy. The ED threshold level can be set by the manufacturer.

- *CCA mode 2.* Mode 2 uses only the CS result, and the channel is considered busy only if the occupying signal is compliant with the PHY of the device that is performing the CCA.

- *CCA mode 3.* This mode is a logical combination (*AND/OR*) of mode 1 and mode 2. In other words, in mode 3, the PHY can use one the following as an indication of a busy channel:

  - The detected energy level is above the threshold *and* a compliant carrier is sensed.

  - The detected energy level is above the threshold *or* a compliant carrier is sensed.

The CCA mode that the device will use is stored as a PHY attribute (*phyCCAMode*) in the PHY PAN Information Base (PHY-PIB). The PHY-PIB is reviewed in the following section.

3.2.6 The PHY Constants and Attributes

The constants define the characteristics such as the maximum size of a frame or the duration of an event. Each layer of the protocol has its own constants. The PHY has only two constants, shown in Table 3.2. The PHY constant of *aMaxPHYPacketSize* indicates that the PHY Service Data Unit (PSDU) cannot exceed 127 octets. The turnaround time is the time a transceiver needs to switch from transmitting (TX) to receiving (RX), and

Table 3.2: PHY Constants

Constant	Description	Value
aMaxPHYPacketSize	The maximum allowed PSDU size (in octets)	127
aTurnaroundTime	The maximum allowed RX-to-TX or TX-to-RX turnaround time (in symbol periods)	12

vice versa. Based on the *aTurnaroundTime* constant, a transceiver must complete its transition in fewer than 12 symbols.

In the PHY and MAC protocol layers, all the constants have a general prefix of *a*. In the NWK and APL layers, in contrast, the prefixes for the constants are *nwkc* and *apsc*, respectively. The constants cannot be changed during operation.

The attributes are the variables that may change during operation. The PHY attributes are contained in the PHY PAN Information Base (PHY-PIB). These attributes are required to manage the PHY services. The list of PHY-PIB attributes is provided in Table 3.3. The attributes marked with a dagger (†) are read-only attributes. The higher layers can read the read-only attributes, but only PHY can change them. The attributes marked with an asterisk (*) have specific bits that are read-only. The bits that are not marked as read-only can be read or written by the next higher layer. Only the PHY can change the read-only bits. The roles of these attributes are clarified in the remainder of this chapter.

3.2.7   PHY Services

The PHY layer provides two types of services: the PHY data service and the PHY management service. The PHY data service enables PHY Protocol Data Unit (PPDU) transmission and reception across a radio channel. The PHY includes a management entity called the Physical Layer Management Entity (PLME), shown in Figure 3.2. The PHY management functions can be invoked from the PLME. The PHY data service is accessed through the PHY Data SAP (PD-SAP). The PHY management service is accessed through the PLME-SAP. The PLME also maintains the PHY PAN Information Base (PIB).

Table 3.3: PHY PIB Attributes

Attribute	Description
phyCurrentChannel	The frequency channel of operation
phyChannelsSupported†	The array of the available and unavailable channels
*phyTransmitPower**	The transmitter output power in dBm
phyCCAMode	The CCA mode of operation (1–3)
phyCurrentPage	The current PHY channel page
phyMaxFrameDuration†	The maximum number of symbols in a frame (55, 212, 266, 1064)
phySHRDuration†	The duration of the synchronization header (SHR) (3, 7, 10, 40)
phySymbolsPerOctet†	The number of symbols per octet for the current PHY (0.4, 1.6, 2, 8)

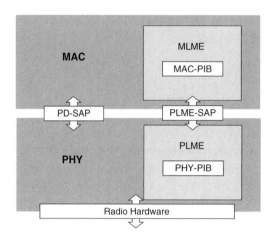

Figure 3.2: The IEEE 802.15.4 PHY Reference Model Interfacing the MAC Layer

3.2.7.1   PHY Data Service

The data that needs to be transmitted is always provided as a MAC Protocol Data Unit (MPDU). The local MAC generates the request for transmission and provides the MPDU. The PHY attempts the transmission and reports the result of the attempt (successful or unsuccessful) to the MAC. The reasons for an unsuccessful transmission attempt can be any of the following:

- Radio transceiver is disabled.

- Radio transceiver is in receive mode. Radio cannot transmit and receive simultaneously.

- Radio transceiver is busy (already engaged in transmitting).

When the data is received by the radio transceiver, the PHY notifies the MAC layer of arrival of an MPDU. The PHY not only provides the MPDU to the MAC layer, it also delivers the LQI information.

Figure 3.3 reviews the data transfer steps from the application layer of one device to another. The data does not always come from the application layer. The data, for example, can be generated by the MAC layer without engaging the next higher layer. In the scenario shown in Figure 3.3, the data is provided by either the ZigBee Device Objects (ZDO) or an application object to the Application Support sublayer (APS). In the

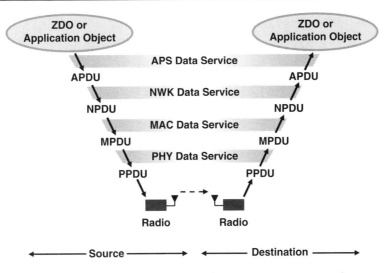

Figure 3.3: Data Transfer Service Between Two Devices

transmitting device, each layer adds its own header and footer (if applicable) to the data unit (DU) and then passes the result to the next lower layer.

The DU in each layer is known by the name of the layer. In the APS and NWK layers, the DU is called the APS Protocol Data Unit (APDU) and the NWK Protocol Data Unit (NPDU), respectively. The PHY data service receives a MAC Protocol Data Unit (MPDU) and creates the PHY Protocol Data Unit (PPDU) that will be transmitted by the radio.

On the receiver side, the data is passed upward from one layer to the next higher layer and the header and footers are removed until the DU reaches the intended layer at the destination.

3.2.7.2 PHY Management Service

The Physical Layer Management Entity SAP (PLME-SAP) shown in Figure 3.2 is utilized to transport commands between the MAC Layer Management Entity (MLME) and the PLME. The services provided through the PLME-SAP are:

- Clear channel assessment (CCA)

- Energy detection (ED)

- Enabling and disabling the radio transceiver

- Obtaining information from the PHY-PIB

- Setting the value of a PHY-PIB attribute

3.2.7.2.1 Clear Channel Assessment The MLME requests that the PLME perform a CCA whenever the CSMA-CA requires an assessment of the channel. The result of a CCA can be any of the following:

- The transceiver is disabled and therefore no CCA is performed.

- The channel is available (idle) and can be used for transmission.

- The channel or transceiver is busy:

  - The channel is busy (another device is using this frequency channel).

  - The transceiver is busy transmitting and therefore no CCA is performed.

The PLME does not distinguish between a busy channel and a busy transceiver and delivers the same busy status to the MLME in both cases.

3.2.7.2.2 Energy Detection The ED request is generated by the MLME and issued to the PLME. If the ED measurement is completed successfully, the energy level is reported back to MLME. A disabled radio or a transceiver engaged in transmission will cause the ED request to fail.

3.2.7.2.3 Enabling and Disabling the Radio Transceiver The MLME can request that the PLME put the transceiver in one of the three main states: transceiver disabled, transmitter enabled, and receiver enabled.

3.2.7.2.4 Obtaining Information from the PHY-PIB The PLME can read the value of any PHY attribute in the PHY-PIB and provide it to the MLME. The request to read a PHY attribute is always issued by the MLME.

3.2.7.2.5 Setting the Value of a PHY-PIB Attribute The read-only PHY attributes can be changed only by the PHY. However, for all other attributes, the MLME can request that the PLME set the PHY-PIB attribute to a given value.

3.2.8 The Service Primitives

The IEEE 802.15.4 and ZigBee standards use the concept of primitives to describe the services that a layer provides to the service user in the next higher layer. The communications between the adjacent protocol layers are managed by calling functions or passing messages, called *primitives*, between the layers.

Although each layer has a different role in the overall system, the way each layer operates has some similarities to other layers. For example, the PHY, the MAC, and the NWK layers provide data service to their next higher layers. In all three layers, the mechanism to request a data unit to be transmitted is similar: The next higher layer uses the Data Service Access Point (Data SAP) of the layer below to request the data to be transmitted. If the data transmission is successful, the lower layer provides a confirmation to the upper layer about the status of the transmission.

Because of these similarities, service primitives are found to be useful in describing the capabilities of each standard protocol layer. Each primitive specifies the action to be performed or provides the result of a previously requested action. A primitive may also carry the parameters needed to perform its task.

Figure 3.4 shows this generic way of describing the services a layer provides to its next higher layer. As shown in this figure, there are four generic types of service: request, indication, response, and confirm. In other words, all the services discussed in the IEEE 802.15.4 and ZigBee standards fall into one of these four categories. The service primitives are described in the following formats:

<The primitive>.request

<The primitive>.indication

<The primitive>.response

<The primitive>.confirm

The *request* service primitive (or simply *request primitive*) is generated by the layer N + 1 to the layer N to request a service to be initiated.

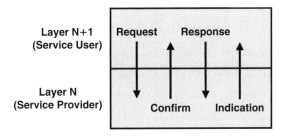

Figure 3.4: The Service Primitive Concept

For example, in the PHY data service, the PHY Data request (*PD-Data.request*) primitive is generated by the MAC layer to the PHY requesting an MPDU to be transmitted.

The *indication* primitive is generated by the layer N to the service user (i.e., the next higher layer), indicating an event that is important for the layer N + 1. For instance, when the PHY receives data from another device in the network that needs to be passed to the MAC layer, the PHY uses the *PD-Data.indication* primitive to deliver the data information to the MAC.

If the *indication* primitive requires a response, the *response* primitive is passed from the service user to layer N. The PHY and NWK layers do not have any *response* primitive. The MAC and the APL layers contain *response* primitives.

The *confirm* primitive is used by the layer N to confirm completion of the service layer N + 1 requested by passing down a *request* primitive. The *PD-DATA.confirm* primitive is generated by the PHY entity and issued to its MAC sublayer entity in response to a *PD-DATA.request* primitive. In the confirmation, the PHY informs the MAC whether the transmission was successful.

This brief overview of the service primitive concept is provided here in case a reader wants to pursue further details of a specific service directly from the IEEE 802.15.4 and ZigBee standards documents. This book uses the service primitives as a convenient way of describing the services provided by each layer but avoids getting into the details of the service primitives whenever possible.

3.2.9   PHY Packet Format

The PHY Protocol Data Unit (PPDU) format is shown in Figure 3.5. The PPDU consists of three components: the Synchronization header (SHR), the PHY header (PHR), and the PHY payload.

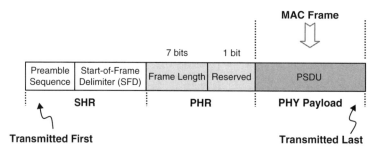

Figure 3.5: PPDU Format

The SHR enables the receiver to synchronize and lock into the bit stream. The PHR contains frame length information. The PHY payload is provided by upper layers and includes data or commands that need to be transmitted to another device.

The SHR consists of a preamble and a start-of-frame delimiter (SFD). The preamble field is used by the receiver to obtain chip and symbol synchronization. The bits in the preamble field in all PHYs, except for the ASK PHYs, are binary zeros. The preamble in an 868 MHz ASK PHY is generated by repeating sequence 0 of Table A.1 twice (see Appendix A). The duration of this preamble is 160 μs. In a 915 MHz ASK PHY, the sequence 0 of Table A.2 is repeated six times and takes 120 μs. The lengths and durations of the preambles in all PHY options are listed in Table 3.4.

The SFD field indicates the end of the SHR and start of the PHR. The SFD for 868 MHz and 915 MHz ASK PHYs is the inverted sequence 0 of Tables A.1 and A.2, respectively. For all other PHYs, the SFD is an 8-bit field shown in Table 3.5. The lengths of SFD fields are provided in Table 3.6.

The next field in a PHY packet is the frame length, which specifies the total number of octets in the PHY payload (PSDU). The PSDU length can be any value from 0 to 127 octets (see Table 3.2, PHY constants). But practically, based on IEEE 802.15.4-2006,

Table 3.4: Preamble Field Lengths and Durations

PHY Option	Length		Duration (μs)
868 MHz BPSK	4 octets	32 symbol	1600
915 MHz BPSK	4 octets	32 symbol	800
868 MHz ASK	5 octets	2 symbol	160
915 MHz ASK	3.75 octets	6 symbol	120
868 MHz O-QPSK	4 octets	8 symbol	320
915 MHz O-QPSK	4 octets	8 symbol	128
2.4 GHz O-QPSK	4 octets	8 symbol	128

Table 3.5: SFD Field Format (Except for ASK PHYs)

Bits	0	1	2	3	4	5	6	7
Values	1	1	1	0	0	1	0	1

Table 3.6: SFD Field Lengths

PHY Option	Length	
868 MHz BPSK	1 octets	8 symbol
915 MHz BPSK	1 octets	8 symbol
868 MHz ASK	2.5 octets	1 symbol
915 MHz ASK	0.625 octets	1 symbol
868 MHz O-QPSK	1 octets	2 symbol
915 MHz O-QPSK	1 octets	2 symbol
2.4 GHz O-QPSK	1 octets	2 symbol

Table 3.7: Frame Length Values

Frame Length Values	PHY Payload
0 to 4	Reserved
5	Acknowledgment MPDU
6 to 8	Reserved
9 to *aMaxPHYPacketSize*	Any other MPDU

the PSDU length is either 5 octets for a MAC acknowledgment frame or 9–127 for any other MPDU. The frame length values of 0–4 and 6–8 are reserved for potential future applications (Table 3.7).

The last field is the PHY Service Data Unit (PSDU). The content of the PSDU is provided by the MAC as a MAC frame. In IEEE 802.15.4, the first bit that will be transmitted is the least significant bit (LSB) of the SHR. The most significant bit (MSB) of the last octet of the PHY payload is transmitted last.

3.2.10 Summary of the PHY Layer Responsibilities

The PHY layer is the closest layer to hardware and directly controls and communicates with the radio transceiver. The PHY layer is responsible for the following:

- Activating and deactivating the radio transceiver.

- Transmitting and receiving data.

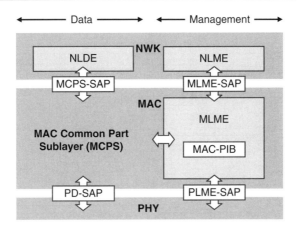

Figure 3.6: The MAC Sublayer Reference Model

- Selecting the channel frequency (the exact frequency in which the transceiver will operate).

- Performing ED. The ED is the task of estimating the signal energy within the frequency band of interest. This estimate is used to understand whether or not a channel is clear and can be used for transmission.

- Performing CCA.

- Generating an LQI. The LQI is an indication of the quality of the data packets received by the receiver. The signal strength can be used as an indication of signal quality.

3.3   IEEE 802.15.4 MAC Layer

The MAC provides the interface between the PHY and the next higher layer above the MAC. In ZigBee wireless networking, this next higher layer is the NWK layer. The IEEE 802.15.4 is not developed specifically for ZigBee applications, and the next higher layer can be any networking protocol layer. In this book, the MAC services are discussed in interfacing with the ZigBee NWK layer.

Figure 3.6 shows the MAC sublayer reference model. The MAC, similar to the PHY, has a management entity called the MAC Layer Management Entity (MLME) that is responsible for the MAC management services. The MLME interacts with its counterpart in the NWK layer (the NWK Layer Management Entity, or NLME).

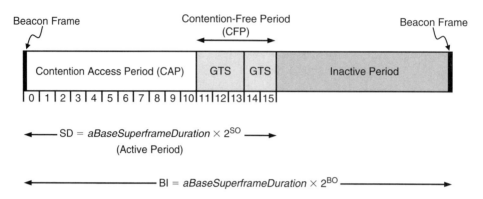

Figure 3.7: The Superframe Structure

The MAC also has its own database, referred to as the MAC PAN Information Base (MAC-PIB). All the MAC constants have a general prefix of *a*. The MAC attributes have a general prefix of *mac*. The size of the MAC-PIB is larger than the PHY-PIB; the table of MAC constants and attributes is included in IEEE 802.15.4 standard document [2].

3.3.1 Beacon-Enabled Operation and Superframe Structure

One of the advantages of a beacon-enabled network is the availability of guaranteed time slots (GTSs). The beacon frames are MAC frames that contain beacon information such as the time interval between the beacons and number of GTSs. The beacon format is reviewed in Section 3.3.5.2.

In beacon-enabled operation, it is possible to use a superframe structure. A superframe, shown in Figure 3.7, is bounded by two beacon frames. The use of a superframe structure is optional in the IEEE 802.15.4 standard. There can be up to three types of periods in a superframe: the contention access period (CAP), the contention-free period (CFP), and the inactive period.

During CAP, all the devices that want to transmit need to use the CSMA-CA mechanism to gain access to a frequency channel. The frequency channel is available equally to all the devices in the same network. The first device that starts using an available channel will keep it to itself until its current transmission is completed. If the device finds the channel busy, it backs off for a random period of time and tries again. This is the most likely mechanism of channel access for the majority of devices in a large network. The MAC command frames must be transmitted during CAP.

There is no guarantee during CAP for any device to be able to use a frequency channel exactly when it needs it. The CFP, in contrast, guarantees a time slot for a specific device and therefore the device does not need to use CSMA-CA for channel access. This is a great option for low-latency applications in which the device cannot afford to wait for a random and potentially long period of time until the channel is available. Using CSMA-CA is not allowed within CFP.

The combination of CAP and CFP is known as the *active period*. The active period is divided into 16 equal time slots. The beacon frame always starts at the beginning of first time slot. There can be up to seven GTSs in CFP. Each GTS can occupy one or more time slots.

A superframe may optionally have an inactive period. The inactive period allows a device to enter power-saving mode. During power-saving mode, the coordinator can turn off its transceiver circuits to conserve battery energy.

The structure of the superframe is defined by the coordinator and configured by the NWK layer using the MLME-START.request primitive. The beacon interval (BI), which is the time duration between two consecutive beacons, is determined by the values of the *macBeaconOrder* (BO) attribute and the *aBaseSuperframeDuration* constant using the equation:

$$BI = aBaseSuperframeDuration \times 2^{BO} \text{ (Symbols)}$$

For example, given *aBaseSuperframeDuration* of 960 symbols and *macBeaconOrder* of 2, the beacon interval will be 3840 symbols. (The MAC constants and attributes are provided in the IEEE 802.15.4 standard document [2].) The *macBeaconOrder* can have any value from 0 to 14 in a beacon-enabled network. If the value of *macBeaconOrder* is set to 15, the network is considered nonbeacon-enabled and none of the superframe discussions will apply.

Similarly, the length of the active period of the superframe, known as the superframe duration (SD), is calculated from the following equation:

$$SD = aBaseSuperframeDuration \times 2^{SO} \text{(Symbols)}$$

where *SO* is the value of *macSuperframeOrder* attribute. The superframe duration cannot exceed the beacon interval; therefore, the value of *SO* is always less than or equal to BO.

In a nonbeacon-enabled network (i.e., where *macBeaconOrder* is equal to 15), the coordinator does not transmit beacons unless it receives a beacon request command

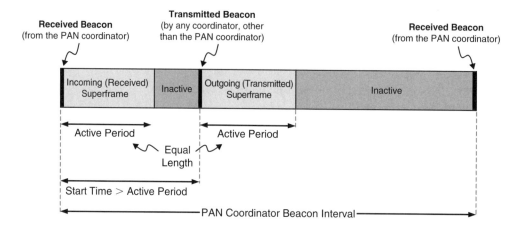

Figure 3.8: The Incoming and Outgoing Superframes' Timing

from a device in its network. The beacon request command is used by a device to locate the coordinator. The format of the beacon request command is provided in Section 3.3.5.5.5. A PAN coordinator in a nonbeacon-enabled network sets the value of *macSuperframeOrder* to 15.

In a beacon-enabled network, any coordinator, in addition to the PAN coordinator, has the option to transmit beacons and create its own superframe. Figure 3.8 shows the required timing when both the PAN coordinator and another coordinator within the same network are transmitting beacons. The coordinator can start transmitting its beacon only during the inactive period of the PAN coordinator superframe. The PAN coordinator beacon is referred to as the *received beacon*. The beacon of any other coordinator is known as the *transmitted beacon*. The active periods of both superframes must have equal lengths. A coordinator, other than a PAN coordinator, only transmits a beacon to specify the start of its superframe, and the end of the superframe can be the same as the end of the PAN coordinator superframe.

If a device does not use its GTS for an extended period of time, its GTS will expire and the coordinator can assign that specific GTS to a different device. The inactive period that will result in GTS expiration is always an integer multiple of twice the superframe length. The value of this integer multiple (*n*) depends on the *macBeaconOrder*:

$$n = 2^{(8-macBeaconOrder)} \quad \text{if } 0 \leq macBeaconOrder \leq 8$$

$$n = 1 \quad \text{if } 8 \leq macBeaconOrder \leq 14$$

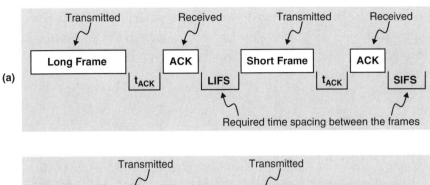

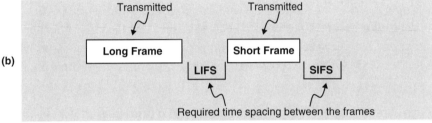

Figure 3.9: The Interframe Spacing (IFS) in (a) Acknowledged Transmission and (b) Unacknowledged Transmission

For example, if a device with *macBeaconOrder* of 7 does not use its GTS in four consecutive superframes, its GTS will expire.

3.3.2 The Interframe Spacing

In transmitting data from one device to another, the transmitting device must wait briefly between its successive transmitted frames to allow the recipient device process a received frame before the next frame arrives. This is known as *interframe spacing* (IFS). The length of IFS depends on the transmitted frame size. The MPDUs with sizes of less than or equal to *aMaxSIFSFramesSize* (a MAC constant with default value of 18 octets) are considered short frames. A long frame is an MPDU with a size that exceeds *aMaxSIFSFramesSize* octets.

The waiting period after a short frame is referred to as *short IFS* (SIFS). The minimum value of SIFS is *macMinSIFSPeriod*. Similarly, a long frame is followed by a *long IFS* (LIFS) with minimum length of *macMinLIFSPeriod*. The values of *macMinSIFSPeriod* and *macMinLIFSPeriod* are 12 and 40 symbols, correspondingly.

Figure 3.9 shows the interframe spacing for two scenarios. In the first one, the message is acknowledged and the wait time between the acknowledgment frame and the next frame is LIFS or SIFS, depending on the frame length. The time period from transmitting a frame and reception of the acknowledgment frame is shown as t_{ACK}. If no acknowledgment is required, the minimum interframe spacing starts from the moment the frame is transmitted (Figure 3.9b).

3.3.3   CSMA-CA

The channel access mechanism supported by the IEEE 802.15.4 MAC is Carrier Sense Multiple Access with Collision Avoidance (CSMA-CA). In CSMA-CA, whenever a device wants to transmit, it performs a CCA to ensure that the channel is not in use by any other device. Then the device starts transmitting its own signal. A brief overview of the CSMA-CA was provided in Chapter 1. This section provides further details.

In addition to transmitting beacons, there are two more occasions on which a device accesses the channel without using the CSMA-CA algorithm:

- The channel access during the contention-free period (CFP).

- Transmitting immediately after acknowledging a data request command. In other words, if a device requests data from a coordinator, the coordinator transmits the acknowledgment followed immediately by the data without performing CSMA-CA between these two transmissions, even during the contention access period (CAP).

There are two types of CSMA-CA: slotted and unslotted. *Slotted CSMA-CA* is referred to as performing CSMA-CA while there is a superframe structure in place. A superframe divides the active period into 16 equal time slots. The back-off periods of the CSMA-CA algorithm need to be aligned to specific time slots discussed below. The *unslotted CSMA-CA* algorithm is used when there is no superframe structure; consequently, no back-off slot alignment is necessary. A nonbeacon-enabled network always uses the unslotted CSMA-CA algorithm for channel access.

If the CCA indicates a busy channel, the device will back off for a random period of time and then try again. This random back-off period in both slotted and unslotted CSMA-CA is an integer multiple of the unit back-off period. The unit back-off period is equal to *aUnitBackoffPeriod* (a MAC constant) symbols.

Figure 3.10 is the flowchart of the CSMA-CA algorithm. In the first step of the algorithm, a decision is made to use either slotted or unslotted CSMA-CA. Three variables are used

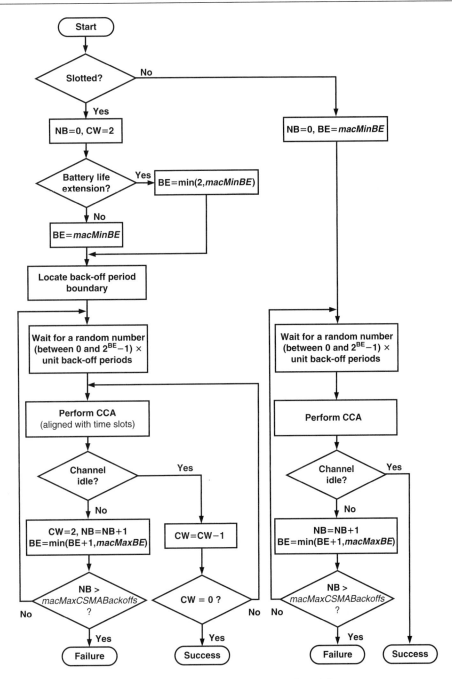

Figure 3.10: The CSMA-CA Algorithm

in the CSMA-CA algorithm: the back-off exponent (BE), the number of back-offs (NB), and the contention window (CW) length.

Every time the algorithm faces a busy channel, it backs off for a random period of time and the BE determines the allowed range for this random period.

This random back-off is any integer number between 0 to $(2^{BE}-1)$ multiplied by the unit back-off period:

$$\text{Back-off} = (\text{A random integer number between 0 to } 2^{BE}-1)$$
$$\times aUnitBackoffPeriod$$

The initial value of BE is equal to *macMinBE* in an unslotted CSMA-CA channel access. In a slotted CSMA-CA, the choice of the battery life extension (BLE) option in the superframe structure affects the value of BE. If the BLE option is active, the coordinator turns off its receiver after a period equal to *macBattLifeExtPeriods* following the transmission of a beacon frame, to conserve energy. In this case, the range for the back-off period is limited to the lesser of 2 and the value of *macMinBE*:

$$BE = \min(2, macMinBE)$$

If the BLE option is not selected, the coordinator is active during the entire CAP, and the value of BE is equal to *macMinBE* (similar to the unslotted CSMA-CA). The value of BE is incremented every time a CCA is performed and the channel is busy. But the value of BE cannot exceed *macMaxBE*.

The NB is a counter that keeps track of the number of times the device backs off and retries the channel access mechanism. At the beginning of the algorithm, NB is equal to zero, and every time the device has to back off due to facing a busy channel, BE is incremented once. If NB reaches *macMaxCSMABackoffs* and still the channel is not accessed successfully, the CSMA-CA algorithm quits and reports channel access failure to the NWK layer.

The contention window (CW) variable determines the number of back-off periods that the channel must be available before starting to transmit. For example, if CW is equal to 2, the device starts transmitting only after two consecutive back-offs resulted in an available (idle) channel. The CW is used only in the *slotted* CSMA-CA algorithm.

If the transmission cannot be completed during the allowed time, the MAC should wait until the start of the next CAP and try the CSMA-CA channel access algorithm again.

3.3.3.1 Hidden and Exposed Node Problems

One of the weaknesses of the CSMA-CA algorithm is the hidden node (terminal) problem [4]. Consider the example shown in Figure 3.11a, where nodes A and C are placed too far from each other to be able to receive each other's signals. However, both nodes A and C can communicate with node B. In each node, the signal power decreases as the distance from the antenna is increased. When node C is transmitting, the signal energy level at the node A location is so weak that the node A energy detection mechanism does not detect the presence of another signal and can declare the frequency channel available (idle). Similarly, node C cannot detect the presence of node A signals, either. Now, if both node A and node C simultaneously decide to transmit packets to node B using the same frequency channel, they may find the channel available at the same time and start transmitting the packets concurrently. This will create a collision of packets in node B.

One way to overcome this issue is changing the location of the nodes or increasing the output power of the hidden nodes to ensure that nodes A and C can detect each other's signals. At the software level, the IEEE 802.15.4 MAC does little to help resolve the hidden node issue. For example, the IEEE 802.15.4 does not currently support the request-to-send/clear-to-send (RTS/CTS) handshake mechanism, which is used in the IEEE 802.11 as part of the effort to overcome the hidden node problem.

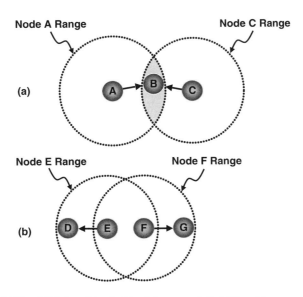

Figure 3.11: (a) The Hidden Node Problem and (b) The Exposed Node Problem

Another related CSMA-CA weakness is the exposed node (terminal) problem. In Figure 3.11b, node E intends to transmit a message to node D while node F is transmitting a message to node G. Node D is outside the radio influence of node F; therefore, node E and node F can concurrently transmit without any collision issue. But CSMA-CA will prevent node E from transmitting because node E is in the radio influence of node F and the CCA performed by node E will consider the channel busy while node F is transmitting. This unnecessary prevention is referred to as the *exposed node problem*. Changing the location of the nodes, reducing the output power of the nodes to the minimum required for a reliable communication, and using the RTS/CTS handshake mechanism are suggested methods for overcoming the exposed node problem.

3.3.4 MAC Services

The MAC layer provides two types of service: the MAC data service and the MAC management service. The MAC data service is accessed by the NWK Layer Data Entity (NLDE) through the MAC Common Part Sublayer SAP (MCPS-SAP). The MAC management service is accessed through MLME-SAP.

A full-function device (FFD) must implement the entire MAC *data* service, but there are some capabilities in the MAC *management* service that are optional for FFDs. There are a number of capabilities in both MAC data and management services that are required for FFDs but are optional for reduced function devices (RFDs). The optional capabilities (service primitives) for RFDs are marked with a diamond (♦) in the IEEE 802.15.4 standard document [2]. The capabilities that are optional for both FFD and RFD are marked with an asterisk (*).

3.3.4.1 The MAC Data Service

The MAC provides data service to the NWK layer. The data that needs to be transmitted is provided as the NPDU. The NPDU is placed in the MAC payload, which is called the MSDU. The NWK generates the request for the data transmission through MCPS-SAP and provides the NPDU.

To keep track of different MSDUs in a device, each MSDU is associated with a unique MSDU handle (*msduhandle*). The *msduhandle* is an integer number that identifies the MSDU. For example, if an MSDU needs to be purged from the transaction queue, the MAC sublayer will attempt to find the corresponding *msduehandle* in the queue.

The data sequence number (DSN) may be used as *msduhandle*. The DSN is a MAC attribute stored in the MAC-PIB (*macDSN*). The initial value of the *macDSN* is a random number. Each time a data frame or a MAC command frame is generated, the MAC sublayer copies the value of *macDSN* into the outgoing frame and increments the *macDSN* by one.

There are three options for data transmission:

- *Acknowledged or unacknowledged transmission.* In an acknowledged transmission, the transmitting device requests that the data recipient device send an acknowledgment frame back if the data is received successfully. In unacknowledged transmission, the data recipient does not send an acknowledgment back. In general, sending an acknowledgment is optional and the data recipient device does not send an acknowledgment back unless requested to do so by the message sender.

- *Transmission during GTS or CAP.* In a nonbeacon-enabled network, this transmission option is always CAP because there is no GTS in a nonbeacon-enabled network.

- *Direct or indirect transmission.* As the name implies, in indirect transmission, the data is not transmitted directly to the recipient device. Instead, in a beacon-enabled network, the data can be stored in a coordinator and the recipient device is notified that there is data pending for it in the coordinator. This notification is part of the beacon message transmitted on a regular basis. After receiving the notification, the device sends a data request to the coordinator asking for the data to be transmitted to it. Only a coordinator can manage indirect transmission.

The MSDU can be purged from the transaction queue if requested by the NWK layer. The MAC sublayer simply looks for *msduhanle* associated with the MSDU and purges it from the queue if it is not already transmitted. The purge capability is optional for RFDs.

The data service discussed previously was for a device that wants to transmit data. If the device is receiving data, the MAC data service delivers the data to the NWK layer. In addition to the data itself, the LQI measured during the reception of the MPDU and the time at which the data was received (timestamp) are provided to the NWK layer.

3.3.4.2   The MAC Management Service

The MAC management service is accessed through the MAC Layer Management Entity SAP (MLME-SAP). The MAC commands normally include parameters such as the addressing and security fields and report the result of a request in the form of a status to the next higher layer. The status can have several options such as *SUCCESS* or *INVALID*.

3.3.4.2.1   Managing MAC PIB   The MAC layer, similar to the PHY layer, has its own constants and attributes. The MAC attributes are stored in the MAC PAN Information Base (MAC-PIB) and are accessible to the NWK layer.

The NWK layer not only can request the MLME to obtain the value of an attribute from the MAC-PIB, it can also request the value of an attribute from the PHY-PIB. In the latter case, the MLME simply passes down the request to the PLME and notifies the NWK layer upon receiving the results from the PLME.

The NWK layer can request the MLME via the MLME-SAP to set a MAC-PIB or a PHY-PIB attribute to a given value. The NWK layer cannot change the read-only attributes in the MAC or PHY PIBs. The NWK request to change an attribute in the PHY-PIB is passed down by the MLME to the PLME via the PLME-SAP.

3.3.4.2.2   MAC Reset   The NWK layer can request the MLME to reset the MAC sublayer to its initial condition and clear all the internal variables to their default values. This is known as the *MAC reset operation*. The NWK layer also has the option to request all the attributes in the MAC-PIB to be reset to their default values. The MAC uses the PHY management service to disable the transceiver before resetting the internal variables and attributes.

3.3.4.2.3   Device Association and Disassociation   The *association* is the procedure that a device uses to join a network. The MAC layer provides the association procedures as a service to the NWK layer. It is the NWK layer that manages the network formation. In most cases, the device must perform a MAC reset before starting the association procedure.

In this section, the association procedure is explained using service primitives as an example of how service primitives help describe the capabilities of a layer. There are four service primitives for the MAC association procedure:

MLME-Associate.request

MLME-Associate.indication

MLME-Associate.response

MLME-Associate.confirm

The indication and response association primitives are optional for RFDs.

MLME-Associate.request is used by the NWK layer of the device that requests joining a coordinator. This request also provides the list of capabilities of the device that requests joining the network. This list, for example, determines whether the device is an FFD or an RFD. The complete list is reviewed in Section 3.3.5.5.1, where the command format is discussed.

When the MAC layer of the device that wants to join the network receives the association request from its own NWK layer, it passes the command down to the PHY layer as an MPDU. The MPDU becomes the PHY payload and is transmitted by the radio to the coordinator device. When the coordinator response arrives back at the device, the MAC of the device passes a confirmation (*MLME-Associate.confirm*) to the NWK layer regarding the result of the request.

On the coordinator side, when the MAC of the coordinator receives the association request, it uses the *MLME-Associate.indication* primitive to let the NWK layer of the coordinator know about the request. The network layer of the coordinator uses the *MLME-Associate.response* primitive to convey the decision to its own MAC layer.

The coordinator MLME does not directly send the decision back to the unassociated device. Instead, the coordinator MLME uses indirect transmission. Therefore, the data is stored in the coordinator. The unassociated device sends a data request to the coordinator after waiting for a predetermined period of time from the moment the device sent its association request. The length of this required wait time is stored in the device MAC PIB as an attribute (*macResponseWaitTime*).

The association procedure is shown as a sequence chart in Figure 3.12. The MLME-*COMM-STATUS.indication* primitive (reviewed in the next section) provides the transaction status (successful or unsuccessful) to the NWK layer. A device that has successfully joined a network is known as an *associated device*.

Disassociation is a procedure that an associated device uses to notify the coordinator that the device intends to leave the network. The NWK layer of the associated device generates the disassociation request to its own MLME using the *MLME-DISASSOCIATE. request* primitive. This request is then sent to the coordinator using the device PHY data

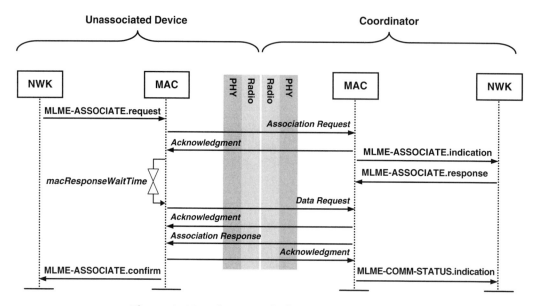

Figure 3.12: The Association Sequence Chart

service. In the disassociation request, the device provides the reason for the request. The reason can be one of the following:

- The coordinator wants the device to leave the PAN.

- The device wants to leave the PAN.

The coordinator MLME analyzes the request and if all the addressing and security fields of the request are valid, the coordinator MLME sends back the confirmation of successful disassociation to the device. This confirmation can be sent back to the device in either a direct or an indirect transmission mechanism. The preferred transmission method is specified in the MLME-DISASSOCIATE.request.

The coordinator MLME uses the *MLME-DISASSOCIATE.indication* primitive to notify its NWK layer regarding the result of the disassociation request received from a device in its network. Figure 3.13 is the sequence chart for the disassociation procedure.

The disassociation request can be initiated by either the device or the coordinator. Figure 3.13a shows the sequence that occurs when the device initiates the disassociation procedure. If the coordinator initiates the disassociation in a beacon-enabled network,

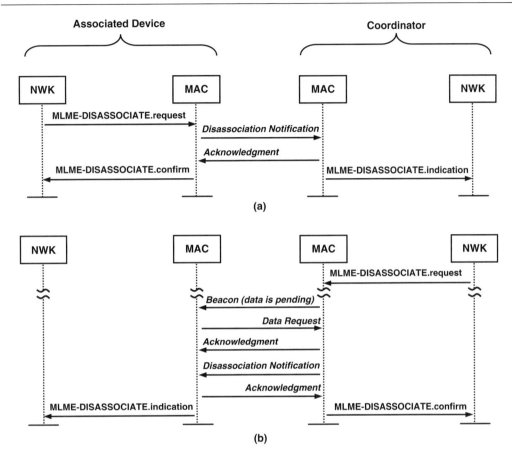

Figure 3.13: The Disassociation Sequence Chart When the Request is Initiated by (a) The Device and (b) The Coordinator

the indirect transmission can be employed. In Figure 3.13b, the coordinator notifies the device that there is data pending for the device as part of the periodically broadcast beacon. After receiving the beacon, the device requests the data and receives the disassociation notification. The *MLME-DISASSOCIATE.confirm* is always used to let the NWK layer of the device that requested the disassociation know the result of the request.

3.3.4.2.4 Communication Status   The MLME uses the *MLME-COMM-STATUS. indication* primitive to provide information such as transmission status to the NWK layer. This primitive is also used by the MLME to report any security-related errors in the incoming packets. If the communication was unsuccessful, the primitive also provides the reason for failure. The unsupported security features or channel access failure are examples of the reasons for unsuccessful communication.

3.3.4.2.5   Enabling and Disabling the Receiver

The NWK layer can request the MLME to enable the receiver for a given fixed period of time. The duration that the receiver will stay on is provided by the NWK layer. The NWK layer also can request to turn off the receiver. These are optional functionalities for both FFDs and RFDs.

The enabling and disabling requests are treated as secondary to other MLME responsibilities. For example, if the MLME has a conflicting responsibility such as transmitting a beacon, the MLME will ignore the NWK layer request of turning on the receiver. The MLME always informs the NWK layer regarding the result of its request to enable or disable the receiver.

3.3.4.2.6   GTS Management

In a beacon-enabled network, there are GTSs that a device can use to transmit without using CSMA-CA. The NWK layer of a device can use the MAC management service to request allocation of a new GTS. If the device already has an assigned GTS and does not need it any longer, the MLME can request the PAN coordinator to deallocate the existing GTS. The NWK layer of the PAN coordinator also can request its own MLME to deallocate an existing GTS allocated to a device in its network.

The PAN coordinator has the option to accept or deny a GTS request. If the PAN coordinator accepts allocating a GTS, the PAN coordinator will include the GTS characteristics, such as its length, in the response. The GTS request primitive is *MLME-GTS.request* and is issued by the NWK layer to the MLME. The MLME communicates the result of a GTS request back to its NWK layer using the *MLME-GTS.confirm* primitive. In a PAN coordinator, the MLME uses the *MLME-GTS.indication* primitive to inform its NWK layer whenever the PAN coordinator allocates or deallocates one of the GTSs based on receiving the request from any device in the network. If the NWK layer has requested the GTS allocation or deallocation, its MLME will notify the NWK layer using the *MLME-GTS.confirm* primitive.

Figure 3.14 shows the sequence chart for the GTS allocation when initiated by a device. Figure 3.15 is the sequence chart for two different deallocation scenarios. Figure 3.15a is the deallocation procedure, when the device initiates the request. In Figure 3.15b, the PAN coordinator is the initiator.

The deallocation of a GTS can leave an unused gap in the CAP. For example, in Figure 3.16, GTS2 is deallocated and time slots 11 to 13 are no longer used by any device. This

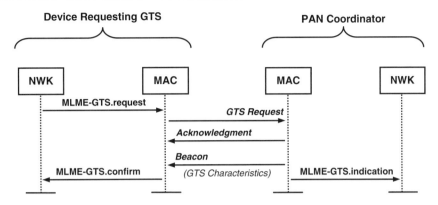

Figure 3.14: The Sequence Chart for GTS Allocation Initiated by a Device

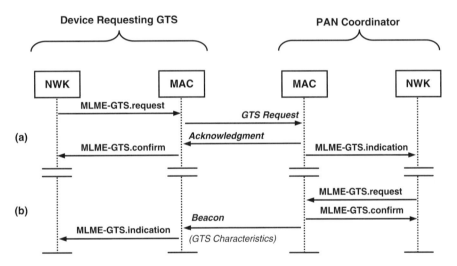

Figure 3.15: The Sequence Chart for GTS Deallocation in Two Different Scenarios: (a) Initiated by a Device and (b) Initiated by the PAN Coordinator

is known as a *fragmented superframe*. To fix this fragmentation issue and increase the CAP, GTS2 is removed and GTS3 is simply *reallocated* to the time slots 12 and 13. As a result of GTS reallocation, the CAP is increased from 9 time slots to 12 time slots.

3.3.4.2.7   Updating Superframe Configuration   In a beacon-enabled network, the NWK layer can request the MLME to start a superframe structure. The NWK layer

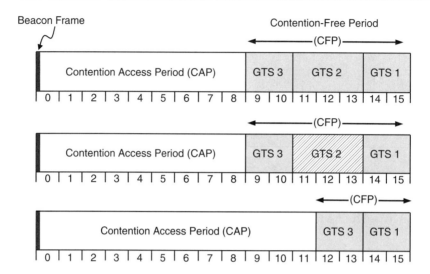

Figure 3.16: The GTS Reallocation to Fix a Fragmented Superframe

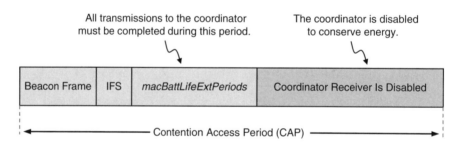

Figure 3.17: If the BLE Option is Selected, the Coordinator Receiver is Active Only for a Limited Time after the IFS Period

provides the necessary parameters, including but not limited to the length of the active period and how often the beacons must be transmitted.

One of the parameters in the superframe configuration is the BLE option (see Figure 3.17). This option allows the beaconing coordinator to turn off its receiver for a period of time equal to *macBattLifeExtPeriods* after transmitting its beacon frame to conserve battery energy. This period is in addition to the required IFS period after transmitting any frame.

If the BLE option is set to false, the beaconing coordinator must keep its receiver active for the entire CAP.

3.3.4.2.8 Orphan Notification A device must be associated with a network to be able to communicate with other devices in the network. A device that was previously associated with a network but has lost its association is considered an *orphaned device*. A device that leaves a network using the disassociation procedure is not considered an orphaned device. If the NWK layer of a device faces repeated communications failures, it may conclude that the device has been orphaned.

For example, if the device transmits a frame that requires an acknowledgment and does not receive the acknowledgment after waiting for *macAckWaitDuration* symbols, the device may repeat the data transmission up to *macMaxFrameRetires*. If still, after *macMaxFrameRetires* attempts, no acknowledgment is received, the device counts this as a *single* communication failure. The application developer decides on the number of communication failures that will be tolerated before declaring the device an orphan.

The NWK layer of an orphaned device can instruct its MLME to perform either one of the following procedures:

- Reset the MAC and then perform the association procedure.

- Perform the orphaned device realignment procedure.

The reset and association procedures were discussed in the preceding sections. The realignment procedure (shown in Figure 3.18) starts with a device sending an orphan notification command to the coordinator. The MLME of the coordinator notifies the NWK layer of the presence of an orphaned device using the *MLME-ORPHAN.indication* primitive.

The NWK layer of the coordinator verifies the address of the orphan device, and in its response (using the *MLME-ORPHAN.response* primitive) the NWK layer confirms whether the device was previously associated with this coordinator.

If the device was previously associated with this coordinator, the MLME of the coordinator sends the realignment command to the orphaned device. The realignment command is used to deliver network settings. When the command is successfully transmitted to the orphaned device, the MLME uses the *MLME-COMM-STATUS. indication* primitive to report its success to the NWK layer.

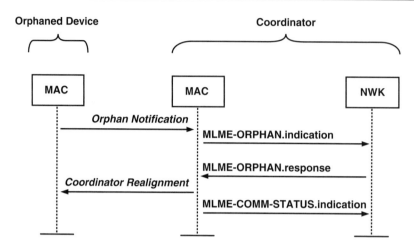

Figure 3.18: The Sequence Chart for Orphan Device Notification

If the device was not associated with this coordinator, the MLME of the coordinator does not perform any action. The orphaned device waits for *macResponseWaitTime* symbols (a MAC attribute) and if it does not receive any realignment command, the orphaned device assumes it is not associated with any coordinator in its range.

3.3.4.2.9   Channel Scanning   The channel-scanning capability is provided as a service of the MAC layer to the NWK layer. The channel scan provides information about the activities within the personal operating space (POS) of the device. There are four types of channel scan:

- *The ED scan.* The energy level in each channel is determined using the PHY energy detection service. This ED scan is optional for RFDs.

- *The orphan scan.* If the device is orphaned, it can search for the PAN to which it is currently associated. In an orphan scan, the MLME sends an orphan notification to the coordinator on each channel and waits for the realignment command from the coordinator. If the device receives the realignment command, it will stop scanning by disabling its receiver. Otherwise the device will continue to the next channel on the list.

- *The active scan.* In this type of scan, the MLME first sends a beacon request command. Then the device enables its receiver to record the information. The active scan can be used by a coordinator that plans to establish its own network to discover all the PAN identifiers used by other networks in its POS and select a unique PAN identifier for its own network. The active scan capability is optional for RFDs.

- *The passive scan.* In the passive scan, in contrast to the active scan, there is no beacon request command transmission. The MLME enables the receiver immediately after it receives the passive scan request and starts recording the received information. The passive scan can be used by an unassociated device to locate a coordinator as part of the association procedure.

3.3.4.2.10   Beacon Notification   When a device receives a beacon, the MLME is required to send the parameters contained in the beacon frame to the NWK layer if the beacon has a payload or autorequest attribute is set to zero. The LQI value and the time the beacon frame was received are also delivered to the NWK layer. The primitive that performs the beacon notification is the *MLME-BEACON-NOTIFY.indication*.

3.3.4.2.11   Synchronizing with a Coordinator   The NWK layer of a device in a beacon-enabled network can request its MLME to synchronize the device with the coordinator using the *MLME-SYNC.request* primitive. The device can choose to locate the beacon only once or can continuously track the beacon. In beacon tracking, the device enables its receiver on a periodic basis just before the expected arrival time of the beacon.

If a device loses its synchronization with the coordinator, the NWK layer requests the MLME to inform the coordinator about the loss of synchronization using the *MLME-SYNC-LOSS.indication* primitive. A device concludes that it has lost its synchronization with the coordinator if it listens for a beacon for a period of *aMaxLostBeacons* (a MAC constant with default value of 4) and does not detect any beacon. Figure 3.19a is the sequence chart showing when the device attempts to locate the beacon but not track it. The beacon-tracking sequence chart is shown in Figure 3.19b.

3.3.4.2.12   Requesting Data from a Coordinator   As discussed earlier in this chapter, the coordinator can use its periodic beacon frame transmission to notify a device in its network that there is data pending for that device in the coordinator. When a device is notified regarding the pending data, the NWK layer of the device requests the MLME (using the *MLME-POLL.request* primitive) to send a data request to the coordinator. This primitive can be used in both beacon-enabled and nonbeacon-enabled networks to request data from a coordinator.

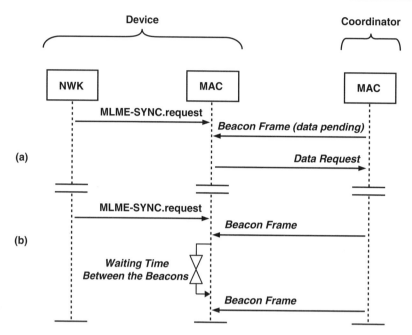

Figure 3.19: Synchronization Sequence Chart Showing When (a) The Device Attempts to Locate the Beacon and (b) The Device is Tracking the Beacon

The device waits for its data for a period of *macMaxFrameTotalWaitTime* symbols. If the MLME does not receive the data during this period, it will notify its NWK layer that no data was received using *MLME-POLL.confirm.*

Figure 3.20 shows two data-polling scenarios. In the first one, the device requests the data and the coordinator sends an acknowledgment back indicating that there is no frame pending (FP = 0). If there is a frame pending (FP = 1), the data follows the acknowledgment frame. The MLME of the data recipient device transfers the data to its NWK layer using the *MCPS-DATA.indication* primitive of the MAC data service.

3.3.5   The MAC Frame Format

The IEEE 802.15.4 defines four general MAC frame structures: the beacon frame, the data frame, the acknowledgment frame, and the MAC command frame. These four frames were reviewed briefly in Chapter 1. This section provides further details of the MAC frame format starting with the general MAC frame.

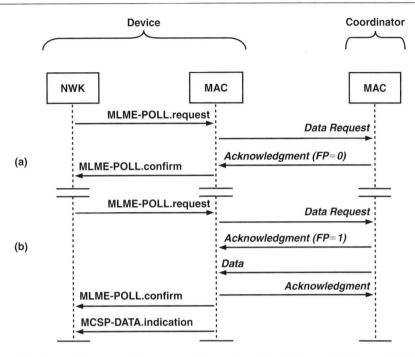

Figure 3.20: Data Request Sequence Chart Showing When (a) There is No Frame Pending (FP=0) and (b) There is a Frame Pending (FP=1)

3.3.5.1 General MAC Frame Format

The general MAC frame is shown in Figure 3.21a and consists of three sections: the MAC header (MHR), the MAC payload, and the MAC footer (MFR). The size of each field is shown in octets.

The first field is the frame control (Figure 3.21b) that defines the frame type (beacon, data, acknowledgment, and MAC command). If the security-enabled subfield is set to 1, this frame has security protection and the auxiliary header will be part of the MAC frame. Otherwise the size of the auxiliary header is zero. The frame pending subfield is used as part of the indirect data transmission method and, if it is set to 1, it means that there is data pending at the transmitting device for the recipient device. If the acknowledgment request subfield is set to 1, the recipient device must send an acknowledgment frame back.

When communicating within the same PAN, the PAN identifier will be the same for both source and destination devices; therefore, it is unnecessary to repeat them both in one

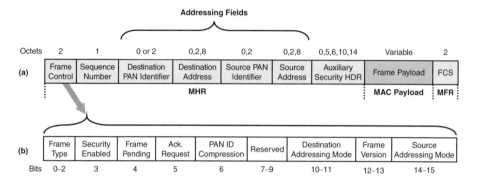

Figure 3.21: (a) General MAC Frame Format and (b) Details of the Frame Control Field

frame. The PAN ID compression subfield helps avoid the unnecessary repeat of the PAN identifier. If the PAN ID compression field is set to 1, only the destination PAN identifier will be included in the frame, and the source PAN identifier is assumed to be the same as the destination.

The destination and source addressing mode subfields determine the addressing mode (16-bit short address or 64-bit extended address). The length of the destination and address fields in the MAC frame depends on the addressing mode. The next subfield is the frame version. The IEEE 802.15.4 standard may be updated over time, and the frame version subfield determines what version of the IEEE 802.15.4 is used to construct the frame.

The next field in the MAC frame, the sequence number, can contain either a beacon sequence number (BSN) or a data sequence number (DSN). The sequence numbering helps distinguishing between various sequences. For example, if two received frames have the same sequence number, it means that the same frame was retransmitted. If the first frame was detected successfully, the second frame (with the same sequence number) can be ignored.

The values of BSN and DSN are stored as MAC PIB attributes (*macBSN* and *macDSN*, correspondingly). The BSN is used only in the beacon frames. The DSN is used in any type of frame other than a beacon frame. A device initializes the *macDSN* value (or *macBSN* value if it is a beacon frame) to a random number and increment it once after each transmission.

The auxiliary security header is an optional field in the MAC frame and contains information such as security level and the type of security keys used to protect the MAC

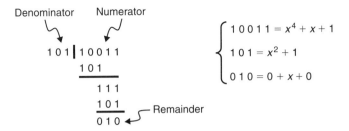

Figure 3.22: A Simple Modulo2 Division Example

frame. (The details of applying security to a frame are discussed in Section 3.6.) The last field in the MAC frame is always the frame check sequence (FCS) field, which the receiver uses to check for any possible error in the received frame. The details of the FCS are provided in following section.

3.3.5.1.1   The Frame Check Sequence   The IEEE 802.15.4 uses 16-bit FCS based on the International Telecommunication Union (ITU) Cyclic Redundancy Check (CRC) to detect possible errors in the data packet [5].

The basic concept of CRC is as follows. In the transmitting device, all the bits in the MHR and the MAC payload are treated as coefficients of a polynomial. This polynomial is then divided by another polynomial, which is known by both the receiver and transmitter. The remainder of this division is called FCS and is added to the end of the frame as the MAC footer (MFR). The recipient device will perform the same division and expects to get the same remainder. If the remainder calculated by the recipient device is not the same as the remainder provided by the transmitting device in MFR, the recipient device will conclude that the frame is received with errors.

A very simple example of polynomial division is shown in Figure 3.22. The division is based on modulo2 binary arithmetic [6]. The numerator is binary number 10011, which is equivalent to $x^4 + x + 1$ polynomial:

$$10011 \rightarrow 1 \times x^4 + 0 \times x^3 + 0 \times x^2 + 1 \times x^1 + 1 \times x^0 = x^4 + x + 1$$

If 10011 is divided by 101, the remainder will be 010.

The FCS in IEEE 802.15.4 is generated by going through the following steps:

- Define the polynomial $M(x)$ to represent the MHR and MAC payload sequence of bits.

- Multiply M(x) by x^{16} to create $M(x) \times x^{16}$.

- Divide the $M(x) \times x^{16}$ polynomial by the following polynomial:

$$G_{16}(x) = x^{16} + x^{12} + x^5 + 1$$

The remainder of this division is placed in the MFR as the frame check sequence.

3.3.5.2 The MAC Beacon Frame Format

The beacon frame (see Figure 3.23) is a special form of the MAC general frame format. The sequence number field contains the current value of *macBSN* (beacon sequence number).

The superframe specification field is part of the MAC payload, and its subfields are shown in Figure 3.23. The beacon order (BO) subfield determines the transmission intervals between the beacons. The relationship between the BO and the BI was reviewed in Section 3.3.1. The superframe order (SO), which was also discussed in Section 3.3.1, specifies the active period of the superframe.

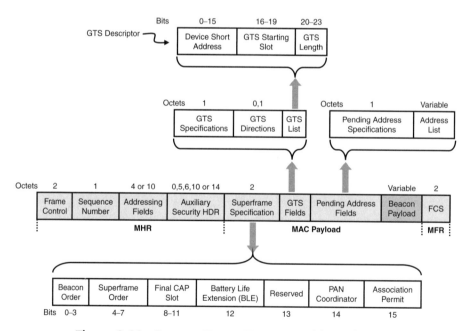

Figure 3.23: Beacon Frame Format and its Subfields

The superframe is divided into 16 equal time slots, and the final CAP slot subfield determines the last time slot of the CAP. If there are any time slots left after the CAP, they are used as GTSs.

If the BLE subfield is set to one, it means that the beaconing device will turn off its receiver after a certain period of time (as discussed in Section 3.3.4.2.7) to conserve its energy.

If this beacon frame is transmitted by the PAN coordinator, the PAN coordinator subfield is set to one. This helps distinguish between a frame received from the PAN coordinator and any other coordinator in the same network. The last bit in the superframe specification field is the association permit. If the association permit is set to zero, it means that the coordinator does not accept association requests at this moment.

The beacon frames are used to establish GTSs. The GTS specification subfield clarifies whether the PAN coordinator currently accepts GTS requests. It also determines the number of GTSs listed in the GTS list field. The GTS direction subfield can be used to set the direction of the GTSs to receive-only or transmit-only modes. The GTS list is the list of all GTSs that are currently maintained (GTS descriptors). Each GTS descriptor contains the short address of the device that will use the GTS, the GTS starting slot, and the GTS length.

The pending address field contains the addresses of all the devices that have data pending at the coordinator. Each device checks for its address in this field and if there is a match, the device will contact the coordinator and request the data be transmitted to it.

The beacon payload is an optional field and is provided by the next higher layer (NWK).

3.3.5.3   The MAC Data Frame Format

The MAC data frame is shown in Figure 3.24a. The data payload is provided by the NWK layer as an MSDU. The MAC data frame is referred to as an MPDU. The sequence number field is equal to the current value of *macDSN* (data sequence number) when the frame is created.

3.3.5.4   The MAC Acknowledgment Frame Format

The MAC acknowledgment frame, shown in Figure 3.24b, is optionally sent by a recipient device to the transmitting device to acknowledge successful reception of a packet. This frame can also be used to let the recipient of the acknowledgment know that there is data pending for that device. This is done by setting the frame-pending subfield of

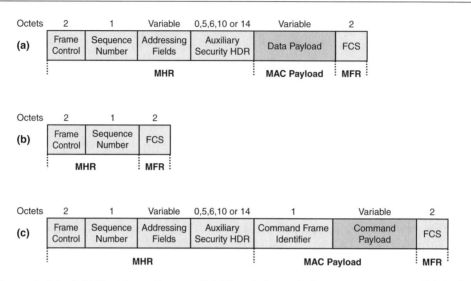

Figure 3.24: (a) The Data Frame, (b) The Acknowledgment Frame, and (c) The MAC Command Frame Formats

the frame control field to one. The sequence number field is equal to the current value of *macDSN*.

3.3.5.5 The MAC Command Frame Format

The MAC command frame (Figure 3.24c) is used to carry the MAC layer commands to the recipient device. The sequence number field is equal to the current value of *macDSN* of the frame being acknowledged.

The list of the MAC commands is provided in Table 3.8. The command frame identifier field determines which one of the commands of Table 3.8 is being executed. The command itself and all the associated parameters are placed in the command payload. (The functional descriptions of all the MAC commands in Table 3.8 are discussed in the following section.) An FFD must be capable of transmitting and receiving all the commands. However, an RFD is required to be capable of receiving or transmitting the commands marked with a checkmark in Table 3.8.

3.3.5.5.1 The Association Request and Response Command Formats   The association command is used as part of the MAC association services (Section 3.3.4.2.3).

Table 3.8: MAC Commands

Command Frame Identifier	Command	RFD	
		TX	RX
00000001	Association request	✓	
00000002	Association response		✓
00000003	Disassociation notification	✓	✓
00000004	Data request	✓	
00000005	PAN ID conflict notification	✓	
00000006	Orphan notification	✓	
00000007	Beacon request		
00000008	Coordinator realignment		✓
00000009	GTS request		

The formats of the association request and response are shown in Figure 3.25. The capability information field of the association request helps answer the following questions regarding the device that requested to join the network:

- Can the device act as a PAN coordinator? The alternate PAN coordinator field is set to zero if the device cannot act as a PAN coordinator.

- Is this an FFD or an RFD? If the device type field is set to one, the device is an FFD.

- Is the device battery powered or connected to a main power source? For battery-powered devices, the power source field is set to zero.

- Does the device keep its receiver in ON mode all the time or turn off the receiver and go to power-saving mode when the device is in idle mode? The receiver is ON when idle field is set to zero if the receiver is turned off during idle mode. The content of this field comes from the MAC attribute *macRxOnWhenIdle*, which is set to FALSE if the receiver is turned off during idle mode.

- Can the device receive and transmit cryptographically protected MAC frames? If it can, the security capability field is set to one.

- Does the device want to have a new address after joining the network? If the device needs a short address, the allocate address field is set to one.

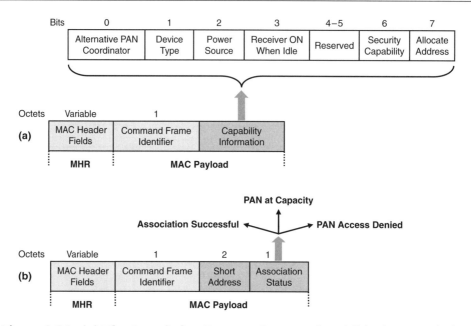

Figure 3.25: (a) The Association Request Command and (b) The Association Response Command

The coordinator uses the association response command to either accept or deny the association request from a device. If the coordinator has reached its capacity, the association response will indicate PAN at capacity status. If the association request is accepted and the device also had asked for a new address, the short address is provided in the short address field of the association response.

3.3.5.5.2   The Disassociation Notification Command Format   The disassociation is a procedure that an associated device uses to notify the coordinator that the device intends to leave the network (Section 3.3.4.2.3). The disassociation frame format is shown in Figure 3.26. A device will leave a network based on either its own decision or the request of the coordinator.

3.3.5.5.3   The Data Request Command Format   The MAC payload in the data request command only contains the command frame identifier (Figure 3.27). This command is used to request data from a coordinator as part of an indirect data transmission method. If the device finds its address in the pending address field of the beacon frame, it knows that there is data pending for it at the coordinator.

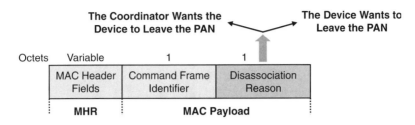

Figure 3.26: MAC Disassociation Notification Command

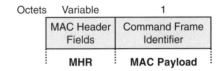

Figure 3.27: The MAC Command Format for Data Request, PAN ID Conflict Notification, Orphan Notification, and Beacon Request

3.3.5.5.4   The PAN ID Conflict Notification Command Format   When a PAN coordinator starts establishing its own network, it will first perform a channel scan to locate other possible networks. Then the PAN coordinator selects a PAN identifier (PAN ID) that is not used by any other nearby network. However, after establishing its network, the PAN coordinator could discover that there is another network with the same PAN ID in its POS. Upon such an event, the PAN coordinator would follow the PAN ID conflict resolution steps to resolve this issue.

There are two ways that a PAN coordinator can notice the presence of another network with the same PAN ID in its vicinity:

- The PAN coordinator receives a beacon frame from another coordinator with the same PAN ID as its own. In a beacon frame, if the PAN coordinator subfield is set to one, it indicates that the beacon is coming from a PAN coordinator.

- A device that is associated with a PAN coordinator notices the presence of another network with the same PAN ID and notifies its own PAN coordinator regarding the conflict using the PAN ID conflict notification command (Figure 3.27). A device concludes that there is a PAN ID conflict if it receives a beacon frame from a PAN coordinator that has the same PAN ID as the PAN that the device is associated with, but the coordinator address does not match what the device has in its MAC PIB. After transmitting the PAN ID conflict notification, if the device receives an acknowledgment from the PAN coordinator, the device MLME will notify its NWK layer, using the *MLME-SYNC-LOSS.indication* primitive, that the synchronization is lost due to PAN ID conflict.

After the PAN coordinator detects a PAN ID conflict via either of these methods, it can perform an active scan to identify the existing PAN IDs and select a new and unique PAN ID. The PAN coordinator then updates the superframe configuration accordingly.

3.3.5.5.5 The Orphan Notification and Beacon Request Command Formats The MAC payload in the orphan notification command only contains the command frame identifier (Figure 3.27) and is used as part of the orphan notification procedure (Section 3.3.4.2.8).

The beacon request command is optional for RFDs and is used to locate all coordinators within devices' POS. The MAC payload contains only the command frame identifier (Figure 3.27).

3.3.5.5.6 The Coordinator Realignment and GTS Request Command Formats The coordinator will use the realignment command (Figure 3.28a) whenever there is a change in the network that requires modification in network settings. For example, if there is a PAN ID conflict, the PAN coordinator may select a new and unique PAN ID. In this case the PAN coordinator uses the realignment command to deliver the new PAN ID to the devices in its network.

The GTS allocation and deallocation (Section 3.3.4.2.6) are performed using the GTS request command (Figure 3.28b). This command is optional for both RFDs and FFDs. The GTS direction field can be set to receive-only or transmit-only. If the characteristic type is set to one, this command is used to allocate a GTS. If it is set to zero, the command is used to deallocate a GTS.

3.3.6 The MAC Promiscuous Mode of Operation

When a device is in the promiscuous mode of operation, it will read all the data it receives, regardless of the intended recipient. This mode is normally used for network monitoring. A device can switch to promiscuous mode of operation by setting the *macPromiscuousMode* attribute to TRUE.

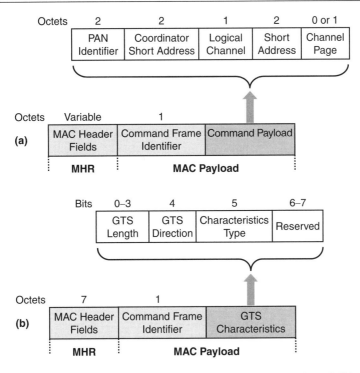

Figure 3.28: (a) The Coordinator Realignment Command and (b) The GTS Request Command

3.3.7   Summary of the MAC Layer Responsibilities

The MAC performs the following duties:

- Generating beacons (if the device is a coordinator)

- Synchronizing the device to the beacons (in a beacon-enabled network)

- Employing the CSMA-CA for channel access

- Managing GTS channel access

- Providing a reliable link between two peer MAC entities (two different devices)

- Providing PAN association and disassociation services

- Providing support for security; the MAC layer is responsible for its own security processing, but the upper layers determine which security level to use

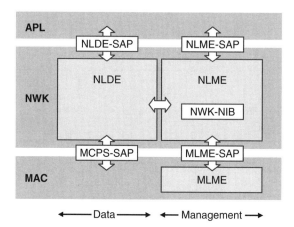

Figure 3.29: The ZigBee Network Layer Interfacing with the MAC and APL

3.4   The ZigBee NWK Layer

The NWK layer (see Figure 3.29) provides two types of services: data and management. The NWK Layer Data Entity (NLDE) is responsible for data transmissions. The data service is accessed through the NLDE Service Access Point (SAP). The NWK management duties are handled by the NWK Layer Management Entity (NLME). The next higher layer can use the NWK management service via NLME-SAP. The NWK layer has its own constants and attributes. All NWK constants start with *nwkc*. The NWK attributes start with *nwk*. The NWK attributes are stored in the Network Information Base (NIB). The list of NWK constants and attributes is provided in the ZigBee standard document [3]. The APL layer can read and modify the NWK attributes using the *NLME-GET* and *NLME-SET* primitives, respectively. A brief summary of ZigBee-Pro additional capabilities is provided in appendix D.

The NWK layer of a ZigBee coordinator assigns a 16-bit *network address* to each device in its network. The ZigBee coordinator, which is also the PAN coordinator, assigns the IEEE 802.15.4 MAC address if a new device that joins its network needs a MAC address. The network address must be the same as the 16-bit IEEE 802.15.4 MAC short address assigned to the device.

The NWK layer limits the distance that a frame is allowed to travel in the network. The distance is defined as the number of hops. A parameter called *radius* is added to each NWK frame to determine the maximum number of hops. For example, if the initial value

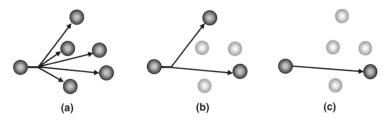

Figure 3.30: (a) Broadcast, (b) Multicast, and (c) Unicast Communication Mechanisms

of the radius is equal to 3, the message will not be relayed more than three times. The value of the radius is decremented every time the message is relayed, and when the radius value becomes zero, the frame will no longer be relayed.

The communication mechanism can be divided into three general categories: broadcast, multicast, and unicast. A *broadcast* message (Figure 3.30) is intended for any device that receives the message. A *multicasting* mechanism delivers the message to a specific group of devices in a network. Finally, *unicast* is used when the message is intended for a single device. A unicast message contains the specific address of the intended device. Unless otherwise specified, unicast is the default mode of communication. The broadcast and multicast modes are explained further in the next two subsections.

3.4.1 Broadcasting

In broadcasting, the message is intended to be received by all devices that are listening to a specific frequency channel, regardless of their address or PAN identifier. Every time a device receives a packet, the device will check the destination address provided in the packet to verify whether the device is the intended recipient of the packet. To broadcast in the IEEE 802.15.4 networks, the short addressing mode is used and the destination address is set to 0xffff. This address will be accepted by all the devices that receive the packet as their own address. The PAN identifier can also be set to 0xffff. The recipient device will accept 0xffff as a valid PAN identifier. The MAC address of 0xffff is known as the *broadcast address*. The PAN identifier of 0xffff is called the *broadcast PAN identifier*.

Although the IEEE 802.15.4 supports the use of a broadcast PAN identifier (i.e., 0xffff) to broadcast a message across multiple networks, the ZigBee standard does not allow broadcasting across multiple networks, and the PAN identifier is always set to the PAN identifier of the ZigBee network instead of 0xffff. The APS sublayer of any device within a network can initiate a broadcast transmission using the NWK layer data service.

In a large network, it would be difficult and unnecessary to expect all the devices that receive a broadcast message to send an acknowledgment back to the message originator. Therefore, when a message is broadcast, the end device is not allowed to acknowledge the successful receipt of the message. Instead, the ZigBee coordinator and ZigBee routers verify whether their neighboring devices have successfully relayed the message. This is known as the *passive acknowledgment* mechanism. In passive acknowledgment, after a device broadcasts a message, it will go into receive mode and wait until the same frame is rebroadcast by any of the neighboring devices. A rebroadcast message is an indication of the fact that a neighbor device has received and relayed the broadcast message successfully.

The ZigBee coordinator and ZigBee routers maintain the record of all the messages that they broadcast in a table called the *broadcast transaction table* (BTT). The record itself is known as the *broadcast transaction record* (BTR) and contains the sequence number and the source address of the broadcast frame. Every ZigBee router is required to be able to buffer at least one frame at the NWK layer. The buffering capability helps in retransmission of the broadcasts. Each BTR is valid for only a limited period of time and will expire after *nwkNetworkBroadcastDeliveryTime* seconds from its creation. An expired BTR can be overwritten if a new BTR is being created and the BTT is full.

If a ZigBee end device does not keep its receiver in ON mode while the device is in idle (i.e., *macRxOnWhenIdle* is set to FALSE), the device neither participates in relaying the broadcast messages nor maintains a BTT. If a ZigBee router with *macRxOnWhenIdle* set to FALSE receives a broadcast message, it will not use the broadcast mechanism. Instead, it will use unicast to relay the message, without any delay, to its neighbors individually. The address field contains the address of the intended device and not the broadcast address.

During broadcasting, the message is relayed by multiple devices and there is a chance of collision due to the hidden node problem (discussed in Section 3.3.3.1). To reduce this chance, the NWK layer requires that before each retransmission the device must wait for a random period of time. This random wait period is called the *broadcast jitter*. The length of the broadcast jitter must be less than the value of the *nwkcMaxBroadcastJitter* attribute in milliseconds.

The sequence chart in Figure 3.31 illustrates the broadcast transaction. The BTT is updated after each device receives a broadcast transmission from a neighbor device. Device A initiates the broadcast and receives the broadcast frame back from device B. When device A receives the broadcast message for the second time, it ignores the frame

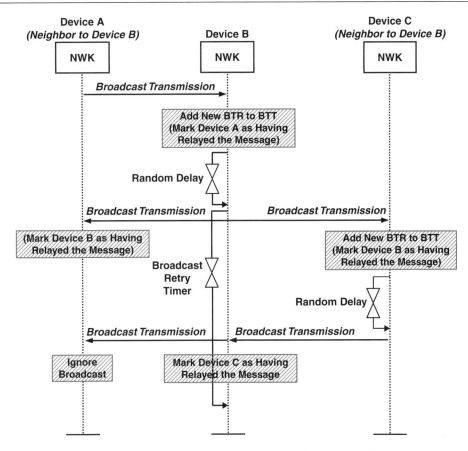

Figure 3.31: Broadcast Transaction Sequence Chart

because device A already received the broadcast frame. Device B broadcasts the frame and waits for a passive acknowledgment (the broadcast frame from device C). If device B waits for a period of *nwkPassiveAckTimeout* seconds and no passive acknowledgment is received, device B will retransmit the frame. In the example shown in Figure 3.31, device B does not retransmit the frame because it has received the passive acknowledgment.

3.4.2 Multicasting

In multicast, the message is delivered to a group of devices within the same network instead of the entire network. For example, in a light control application, a single frame transmitted by a device acting as the switch can turn on or turn off a group of lights in

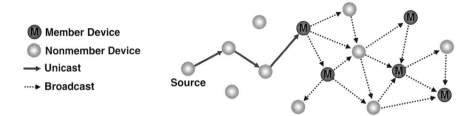

Figure 3.32: Multicasting Initiated by a Nonmember Device

a house. Although it is possible to accomplish the same result with consecutive unicast transmissions to each single light, a single multicast is a more efficient way of delivering the same message to a group of devices.

Each group is identified by a 16-bit multicast group ID. The devices in the same group are known as *group members*. A device can be a member of more than one multicast group. Each device keeps the list of its multicast group memberships in a table called the *multicast table* (*nwkGroupIDTable*).

A device does not have to be a member of a multicast group to be able to use multicasting to reach the members. There are two modes of operation in multicasting: member mode and nonmember mode. In *member mode*, a multicast is initiated by a member device and sent to the members of its multicast group. In *nonmember mode*, a device that is not a member of a multicast group routes the message to a multicast group member, from which the message will be sent to the members of the group.

Figure 3.32 can be used to explain both member mode and nonmember mode multicasting. The NWK data frame itself has a field (the multicast mode field) that clarifies whether this frame is being transmitted by a member device or a nonmember device. In this example, the multicast is initiated by a nonmember device. A device "knows" that it is not a member of the intended multicast group if the multicast group ID included in the payload received from the APL layer does not match any of the device multicast group IDs stored in its multicast table. The initiating device (source device) in Figure 3.32 composes the NWK data frame and sets the multicast mode field of the frame to nonmember mode.

Assuming a route discovery has been performed prior to this transmission, the source device knows the address of the next hop. It means that a route is already established that leads this frame to a member device by means of consecutive unicasts. The source device sets the destination address field equal to the address of the next hop. The next device that receives the frame is also a nonmember device and just passes the frame to the next device using unicast. This process is continued until the frame is received by a device that is a member of the multicast group.

The member device will change the multicast mode field of the received frame to indicate that the message is being sent from a member device. From this point, the message is sent as though it were initiated by a member device. The multicast member devices do not unicast the message. Instead, the message is broadcast by setting the destination address of the frame equal to the broadcast address (0xffff). The member mode multicasts are recorded in BTT as though they were broadcasts. In multicast, in contrast with broadcast, there is no passive acknowledgment.

Any device that participates in broadcasting will wait for a random period of time (less than the value of the *nwkcMaxBroadcastJitter* attribute), then broadcast the frame. This broadcast frame can reach both members and nonmembers (Figure 3.32). Both members and nonmembers will rebroadcast the received frame if their BTT is not full and they have not previously broadcast the same frame.

In multicasting, it is possible to limit the number of times a multicast frame is rebroadcast by nonmember devices. The broadcast frame has a field called *nonmember radius*, which is decremented every time the frame is rebroadcast by a nonmember device. When the nonmember radius field becomes zero, the frame is no longer rebroadcast by nonmember devices.

The only exception is when the nonmember radius field is set to 007, which indicates that there is no limit for the number of times the frame can be rebroadcast. Also, every time a member device rebroadcasts the multicast frame, it will set the nonmember radius field to the maximum nonmember radius allowed. The value of the maximum nonmember radius is also included in the frame that is being multicast.

In ZigBee standard, multicast is used only for data frame transmission, and no command frame is allowed to be transmitted using multicast.

3.4.3   Many-to-One Communication

Figure 3.33 shows a communication scenario in which a single device receives messages from multiple devices within the same network. This is known as *many-to-one* communication. The device that receives the messages is called the *sink*. In ZigBee wireless networking the sink will establish routes to itself from all ZigBee routers and ZigBee coordinators within a given radius.

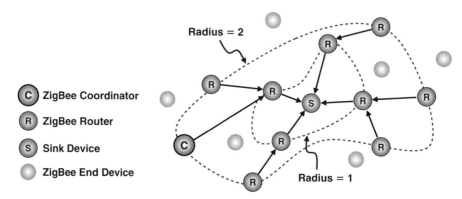

Figure 3.33: The Many-to-One Communication

3.4.4   Hierarchical (Tree) Topology

A tree network, shown in Figure 3.34, starts from a ZigBee coordinator acting as the root of the tree. A ZigBee coordinator or router can act as a parent device and accept association from other devices in the network. A device connected to a parent device is known as a *child device*. The messages intended for a child can be routed through its parent. A ZigBee end device can act only as a child because it lacks the routing capability. The tree topology is also known as a *hierarchical* topology.

The network *depth* is defined as the minimum number of hops required for a frame to reach the ZigBee coordinator if only parent/child links are used. The ZigBee coordinator's direct children have network depth of 1 because they can send a frame to the ZigBee coordinator directly with a single hop. The ZigBee coordinator itself has a depth of zero.

ZigBee standard offers a mechanism to allocate addresses to devices in a tree network. This is known as the *default distributed address allocation*. However, application developers are allowed to use their own address allocation method. When the ZigBee coordinator starts establishing the network, if the *nwkUseTreeAddrAlloc* attribute is set to TRUE the coordinator will use the default distributed addressing scheme. In distributed addressing, the ZigBee coordinator provides each potential parent with a subblock of network addresses. The parent will assign these addresses to its children. The ZigBee coordinator determines the maximum number of children allowed for each parent. If the *nwkUseTreeAddrAlloc* is set to FALSE, the APL layer will provide user-defined addressing to the NWK layer.

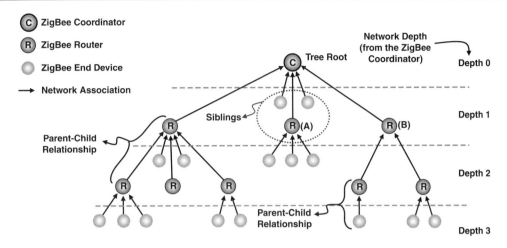

Figure 3.34: The Parent Child Relationship in a Tree Network

The default address allocation simply uses the depth and maximum number of children to allocate the addresses. The parameters that affect the address allocation are the following:

- *Lm.* The network maximum depth (*nwkMaxDepth*).

- *Cm.* The maximum number of children a parent can accept (*nwkMaxChildren*).

- *Rm.* The maximum number of *routing-capable* children a parent can accept (*nwkMaxRouters*).

- *d.* The depth of a device in a network.

Figure 3.35 shows an example of an address allocation, where $Lm = 3$, $Cm = Rm = 2$. The address allocation starts with assigning the zero address (*addr* = 0) to the ZigBee coordinator. To determine the address of the rest of the devices, a simple function (*Cskip(d)*) is introduced:

$$Cskip(d) = \begin{cases} 1 + Cm \times (Lm - d - 1) & , if \ Rm = 1 \\ \dfrac{1 + Cm - Rm - Cm \times Rm^{Lm-d-1}}{1 - Rm} & , \text{otherwise} \end{cases}$$

In the example of Figure 3.35, the value of *Cskip(d)* at each depth is calculated. At each depth, the difference between the address of any two *routing-capable* devices is an integer

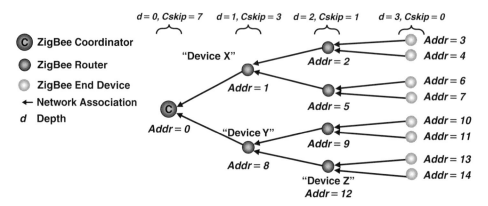

Figure 3.35: Example of Default Address Allocation

multiple of the value of *Cskip* of their parent. For example, in Figure 3.35, the device X address is a single increment of the ZigBee coordinator address (*addr* = 1). The device Y address is the address of device X plus the value of *Cskip* of their parent (*Cskip* = 7). Therefore, the address of device Y is equal to 8 (*addr* = 8). The address of the rest of the routing-capable devices can be determined similarly.

If the calculated value of *Cskip* becomes zero for a device, it means that the device cannot accept any children. In Figure 3.35, the *Cskip* value is equal to zero for any device at the depth of three. Therefore, any device at depth three can act only as an end device. The address of an end device that is not capable of routing is calculated differently. The address of each end device is assigned using the following equation:

$$\text{The } n^{\text{th}} \text{ end device address} = \text{Parent address} + Cskip(d) \times Rm + n$$

For example, in Figure 3.35, the end devices connected to device Z will have the following addresses:

$$\text{First end device address} = 9 + 0 \times 2 + 1 = 10$$
$$\text{Second end device address} = 9 + 0 \times 2 + 2 = 11$$

The *Cskip*(*d*) can be helpful when a device is expected to relay a message toward a destination on behalf of another device. The relaying device needs to know whether the destination device is a descendant of the relaying device. If the relaying device is at depth

d and its address is equal to A, a destination device with a destination address of D is a descendant of the relaying device if the following relationship is true:

$$A < D < A + Cskip(d-1)$$

For example, in Figure 3.35, device Y is at depth one and has an address of 8 ($A = 8$). Then device Y receives a message that needs to be relayed toward the destination address of 11. Device Y uses the following to realize that the destination address is one of its own descendants:

$$8 < 11 < 8 + 7$$

After realizing that the destination is the descendant of a device, the next step is calculating the address of the next hop. If the destination is one of the device children, the address of the next hop will simply be equal to the destination address. If the destination is not a child but it is a descendant, the address of the next hop is calculated from the following equation:

$$\text{Address of the next hop} = A + 1 + int\left(\frac{D - (A+1)}{Cskip(d)}\right) \times Cskip(d)$$

The function *int* returns only the integer part of any number. For instance, *int*(4.95) is equal to 4.

In Figure 3.35, when device Y acts as a router to relay the message to destination address 11, the address of the next hop will be 9:

$$\text{Address of the next hop} = 8 + 1 + int\left(\frac{11 - (8+1)}{3}\right) \times 3 = 9$$

A simple spreadsheet for the distributed address assignment and next-hop address calculation is provided on this book's companion Website.

The source device is the device that initiates transmission of a frame. In a hierarchical topology, a device will relay a frame only if the frame was received along a *valid path*. A path is valid if one of the following conditions is met:

- The frame is received from one of the device children *and* the source device is a descendant of that child.

- The frame is received from the device parent *and* the source device is not a descendant of the device.

The star topology can be considered a special form of a tree network where the only parent in the network is the ZigBee coordinator. In a star network, all children are at the depth of 1 and will communicate directly with the ZigBee coordinator. None of the children act as a router in a star network. The devices in a star topology cannot communicate directly with any device other than the ZigBee coordinator. The address allocation method discussed in this section is applicable to both ZigBee-2006 and ZigBee-2007. The ZigBee-Pro supports stochastic address allocation as well, which is briefly reviewed in appendix D.

3.4.5   Mesh Topology

In a mesh topology, in contrast to the tree topology, there are no hierarchical relationships. Any device in a mesh topology is allowed to attempt to contact any other device either directly or by taking advantage of routing-capable devices to relay the message on behalf of the message originator. In mesh topology, the route from the source device to the destination is created on demand and can be modified if the environment changes. The capability of a mesh network to create and modify routes dynamically increases the reliability of the wireless connections. If, for any reason, the source device cannot communicate with the destination device using a previously established route, the routing-capable devices in the network can cooperate to find an alternative path from the source device to the destination device. This is clarified further in the route discovery and maintenance subsections.

3.4.6   Routing

Routing is the process of selecting the path through which the messages will be relayed to their destination device. The ZigBee coordinator and routers are responsible for discovering and maintaining the routes in a network. A ZigBee end device cannot perform route discovery, and the ZigBee coordinator or a router will perform route discovery on behalf of the end device.

The length (L) of a path is defined as the number of devices in the path. Figure 3.36 shows an example of two paths with the lengths of five ($L = 5$) and seven (L = 7). The connection between two consecutive devices in a path is called a *link*. The links are numbered l_1 to l_4 in Figure 3.36.

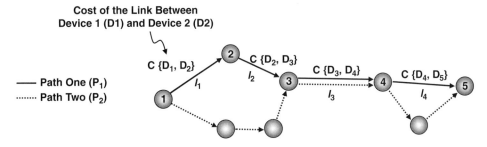

Figure 3.36: The Path-Cost Analysis

The parameters such as link quality, number of hops, and energy conservation considerations can be used to decide on the optimal path for each routing scenario. To simplify this process, each link is associated with a *link cost*. The probability of successful packet delivery on each link will determine the link cost. The lower the probability of successful packet delivery, the higher the cost of the link. The cost of each link is shown as $C\{[D_i,D_{i+1}]\}$ in Figure 3.36.

There are various ways to determine link cost. The ZigBee standard uses the following equation:

$$C\{l\} = \text{The lesser of 7 and } round\left(\frac{1}{p_l^4}\right)$$

$C\{l\}$ is the cost of link l. The probability of successful packet delivery in link l is shown by P_l. The *round* function rounds the number to the nearest integer value. The cost is always an integer between 0 and 7. For example, if P_l is 80%, the cost of the link will be the integer 2:

$$C\{l\} = \text{The lesser of 7 and } round\left(\frac{1}{(0.8)^4}\right) = 2$$

The probability of successful packet delivery (P_l) can be estimated using various methods, and the ZigBee standard allows the implementers to select any method they find most suitable for their application. However, the initial estimate for the probability of successful packet delivery must be based on average LQI. The LQI is recorded for each received packet, indicating the signal energy or the SNR. Generally, the chance of successful reception of a packet increases as the LQI is increased. A simple method to calculate the link

Table 3.9: The Routing Table

Field Name	Size	Description
Destination address	2 bytes	The route will lead to this destination address.
Status	3 bits	The route status can be one of the following: *ACTIVE, DISCOVERY_UNDERWAY, DISCOVERY_FAILD, INACTIVE, VALIDATION_UNDERWAY.*
Many-to-one	1 bit	If the destination has issued a many-to-one route request, this field is set to one.
Route record required	1 bit	If this flag is set, the route taken by the packet will be recorded and delivered to the destination device.
Group ID flag	1 bit	This flag is set if the destination address is a group ID.
Next-hop address	2 bytes	This is the 16-bit network address of the next hop in this route.

cost is to use a lookup table to map different levels of LQI directly to the link cost levels of 0 to 7. This table is usually created based on the average results of several experiments.

To compare different paths, each path is associated with a *path cost*. The path cost ($C\{P\}$) is simply the summation of the costs of the links that form the path:

$$C\{P\} = \sum_{i=1}^{L-1} C\{[D_i, D_{i+1}]\} = \sum_{i=1}^{L-1} C\{l_i\}$$

The route with the lowest path cost will have the best chance of successful delivery of packets.

The ZigBee coordinator and routers create and maintain *routing tables* (Table 3.9). A routing table is used to determine the next hop when routing a message to a particular destination. The status field determines the status of a route. The term *routing table capacity* means that the device is capable of using its routing table to establish a route to a specific destination address.

Another table related to routing is the *route discovery table*, which is used during the discovery of new routes (Table 3.10). The route discovery table contains the path costs, the address of the device that requested the route (source device), and the address of the last device that relayed the request to the current device. The latter will be used to send the result of the route discovery back to the route request originator (source device).

Table 3.10: Route Discovery Table

Field Name	Size	Description
Route request ID	1 byte	The sequence number that identifies a route request. Each route from the source device to the destination device has a unique route request ID.
Source address	2 bytes	The 16-bit network address of the source device. The source device is the route request initiator.
Sender address	2 byte	This is the 16-bit network address of the sender device. The sender device is the device that has sent the route request on behalf of the source device to the current device. This address will be used to send the *route reply* command back to the source device. If the same route request is received from multiple senders, the address of the sender with the lowest overall path cost will be kept here.
Forward cost	1 byte	The accumulated path cost from the source device to the current device. This field is updated when the *route request* command is being sent toward the destination device.
Residual cost	1 byte	The accumulated path cost from the current device to the destination device. This field is updated when the *route reply* command is being sent back to the source device.
Expiration time	2 bytes	The content of the route discovery table expires after a certain period. The initial value of this field is equal to *nwkcRouteDiscoveryTime*.

The content of the route discovery table, in contrast with the routing table, is temporary and expires after *nwkcRouteDiscoveryTime* milliseconds.

A device in a ZigBee network also maintains a *neighbor table*, which contains information about the devices in its transmission range (Table 3.11). This table is updated every time the device receives a packet from one of its neighbors. The neighbor table is useful when the device needs to find a nearby router or rejoin the network. The device also uses the neighbor table when it seeks a new parent. Table 3.11 includes all required fields and some of the optional fields. The list of all optional fields is provided in the ZigBee specification [3].

The routing mechanism for a tree network is called *hierarchical* routing. If the attribute *nwkUseTreeRouting* is set to TRUE, the device is capable of using hierarchical routing.

Table 3.11: The Neighbor Table

Field Name	Description
Extended address	64-bit IEEE 802.15.4 address
Network address	16-bit network address
Device type	ZigBee coordinator, router, or end device
RxOnWhenIdle	If the device keeps its receiver ON during idle mode, this field is set to TRUE
Relationship	This field determines the relationship of the neighbor to the device as parent, child, sibling, previous child, or none of the above
Transmit failure	A high value in this field indicates that many of the previous transmission attempts resulted in failure
LQI	The estimated link quality
Incoming beacon timestamp	The time that the last beacon was received (optional field)
Potential parent	This field determines whether this neighbor is a potential parent (optional field)

3.4.7   Route Discovery

The APL layer can use the *NLME-ROUTE-DISCOVERY.request* primitive to request that the NWK layer discover routes for unicast, multicast, and many-to-one communication. If the route discovery request contains the address of an individual device as the destination address, the NWK layer will perform a unicast route discovery. A unicast route always starts from a single source address and ends at a single destination address. Multicast route discovery will be initiated if the destination address is the 16-bit group ID of a multicast group. Finally, if the APL layer does not provide any destination address, the NWK layer will assume that the APL layer has requested a many-to-one route discovery. The many-to-one routes will place the device that requested the route discovery as the *sink* device.

Only the ZigBee coordinator and routers are capable of carrying out a route discovery request. Also, the ZigBee standard does not allow route discovery for broadcasting. Therefore, if the APL layer issues the *NLME-ROUTE-DISCOVERY.request* primitive to the NWK layer with the destination address equal to the broadcast address (0xffff), the primitive will be treated as an invalid request.

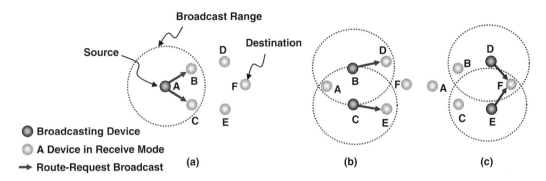

Figure 3.37: Unicast Route Discovery with Device A as the Source and Device F as the Destination

Figure 3.37 shows a unicast route discovery example. In this scenario, device A intends to find a route to device F. Device A starts the route discovery by broadcasting a *route request* command. The *route request* command frame (reviewed in Section 3.4.11) contains the route request identifier, the destination address, and a path-cost field. The route-request identifier is an 8-bit sequence number for route requests and is incremented by 1 every time the NWK layer issues a route request. The path-cost field is used to accumulate the total path cost of each route. Device A will set the value of the path-cost field to zero before broadcasting the *route request* command.

The broadcast command is received by all the devices that are in device A radio range and are listening to the same frequency channel. In Figure 3.37, device B and device C receive the *route request* command. Device A will wait for passive acknowledgment, and if the broadcast was not successful device A will retry the broadcast for *nwkcInitialRREQRetries* times after the initial broadcast. These retries are separated by *nwkcRREQRetryInterval* milliseconds.

If a device that has received the *route request* command is a ZigBee end device, it will ignore the *route request* command because it does not have routing capability. Figure 3.37, for simplicity, shows only the devices located between device A and device F that are either ZigBee coordinators or ZigBee routers. Device B is a ZigBee router, and if device B has routing capacity (routing table is not full), it adds the path cost from device A to device B to the path-cost field of the *routing request* command and broadcasts the route request command. If the device B route discovery table does not contain this *route request* identifier and the source address (device A address), device B will update the route discovery table accordingly.

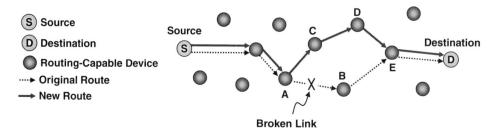

Figure 3.38: Route Repair in a Mesh Network

If the device B routing table is full and the destination address is not in the routing table, assuming that the network is based on hierarchical routing and the *route request* command is received from a valid path, device B will unicast the *route request* command along the tree. But if device B's routing table is full and the route request was not received along a valid path, device B will simply ignore the *route request* command.

Device C is also a ZigBee router and will act similarly to device B when it receives the *route request* command. The broadcasts from devices B and C can reach devices D and E in Figure 3.37. Devices D and E will rebroadcast the route request again, assuming they have routing capacity. Similar to any other broadcasting, each device starts the broadcast with a random delay (broadcast jitter). In route discovery, this delay is a random number between $2 \times nwkcMinRREQJitter$ and $2 \times nwkcMaxRREQJitter$.

Device D, for example, may receive the same *route request* command (initiated by device A) from different relaying devices. Device D will make sure its route discovery table is updated to contain the lowest path cost from the sending device to device D. This will help device D when it is time to communicate back the result of the route discovery toward device A. At that time, device D simply selects its next hop from its route discovery table based on the lowest path cost toward device A.

The consecutive broadcasting is repeated until the *route request* command is received by the intended destination (device F). Device F will use the overall accumulated path cost of each received *route request* command to select the optimum path from device A to device F. Device F will choose either device D or device E as its next hop for transmitting the *route reply* command back to device A. For example, if device D is the next hop, device D will use its route discovery table to find the next hop in relaying the *route reply* command back to device A.

The route from device A (source device) to device F (destination device) is called a *forward route*. The route from device F to device A is considered the *backward route*. The routing can be symmetric or nonsymmetric. When *symmetric routing* is selected, by setting *nwkSymLink* to TRUE the forward and backward routes are identical. In other words, in Figure 3.37, the message is relayed back from device F to device A using the same devices that were used to carry the message from device A to device F. If the value of *nwkSymLink* is set to FALSE, device F needs a separate route discovery to find a route to device A.

Multicast route discovery, similar to unicast route discovery, starts with broadcast of a *route request* command by the source device (initiating device). The source device will create a routing table or update an existing one to reflect the fact that the route discovery is underway. Any device that receives this broadcast and has routing capacity will compare the multicast group ID provided by the broadcast frame with the group IDs in its multicast table to know if the device is a member of the requested multicast group. If the recipient device is not a member of the requested multicast group, it will treat this multicast route discovery similarly to a unicast route discovery, but the destination address is set to the multicast group ID.

If the recipient device is a member of the requested multicast group, it will create or update the route discovery table for this new route discovery request. If this member device already has an entry in its route discovery table corresponding to the source device address and the same route request identifier, the member device will keep the one with the lowest path cost. The member device will send a *route rely* command toward the source device, similar to a unicast route discovery.

3.4.7.1   Source Routing

The ZigBee NWK layer allows the use of source routing. In a source routing mechanism, the frame originator creates the list of all the devices that will act as relays and include them in the NWK frame itself. In this way, when a routing device receives this frame it simply looks up the address of the next node from the relay list included in the frame. This relay list has an index that is incremented every time the frame is relayed to make sure the index always points at the address of the next hop. In other words, in source routing, the device that relays the frame will look up the next-hop address from the relay list instead of its own routing table.

3.4.8   Route Maintenance and Repair

After a route has been discovered and used for a while to relay the messages, there can be incidents in which the route is not capable of relaying a message to the destination.

The reason can be in the network itself (e.g., a router has been removed or turned off) or from the environment (e.g., an object is blocking the wireless connection between two nodes). Considering the amount of effort required to create or repair a route, it is not recommended to start a route repair procedure as soon as a route fails to deliver a message. Instead, it is recommended to keep a counter for the number of times that an outgoing frame has failed due to a link failure. A route repair procedure will start when this counter exceeds the *nwkcRepairThreshold*. It is also possible to record the time that the link has failed. This will help distinguish between temporary and permanent link failures. The application developer will decide on the best method to start the route repair based on the use-case scenario.

If the route error occurred in a many-to-one topology, the device that has faced the link failure will send a route error command frame to a random neighbor. All neighbors are expected to have a routing table to the sink (destination) device. The neighbor will then forward the route error command frame to the sink device.

Figure 3.38 shows an example of route repair in a mesh network. The link between device A and device B has failed and a route repair is required. The procedure for route repair is similar to route discovery except that device A, instead of the original source device, will start the route discovery. Device A will use its own address as the source device when broadcasting a route request command frame to find the new route to the destination device. Both route repair and route discovery use the *route request* command. To distinguish a route repair from a route discovery, the route repair bit subfield in the *route request* command frame is set to one (Section 3.4.11).

In Figure 3.38, after consecutive route request broadcasts, a new route has been established between the source and the destination address. Along the way, the new route may share some of the devices with the original route. If device A, for any reason, cannot establish a new route to the destination, it unicasts a *route error* command frame back to the source device. In this case, the source device has to start a new route discovery instead of route repair to establish routes to the destination address.

3.4.9   The NWK Layer Data Service

The NWK layer receives the data that needs to be transmitted from the APS sublayer in the form of an APS sublayer Protocol Data Unit (APDU). The combination of APDU and the NWK layer header will form the NWK layer Protocol Data Unit (NPDU).

The APS sublayer uses the *NLDE-DATA.request* primitive to request transmission of the data to the APS sublayer of another device in the network. The data transfer request limits the distance that the data is allowed to travel in the network using the radius parameter. The data is accompanied by a sequence number. The sequence number starts from a random number and is incremented once after each data frame is transmitted. On the receiver side, the NWK layer delivers the data frame, LQI, and the data sequence to the APS sublayer using *NLDE-DATA.indication* primitive.

3.4.10 The NWK Layer Management Service

The NWK layer main responsibilities are network formation, joining and leaving a network, route discovery, and route maintenance. The NWK layer routing and route discovery were discussed earlier in this section. The rest of NWK layer management capabilities are reviewed in the following subsections.

3.4.10.1 Network Discovery

The network discovery procedure is used to discover all the networks currently operating in the device POS. The device discovery request is given to the NWK layer by the APL layer. The NWK layer uses the MAC layer channel scanning to discover the presence of other networks. The active scan is the preferred scan method if the device is capable of performing an active scan. Otherwise, the device will perform a passive scan.

The network discovery operation delivers the list of discovered networks, including their PAN identifiers, current frequency channels, and the version of the ZigBee protocol used by the each network to the APL layer. The information also includes the value of beacon order, superframe order of the networks, and their ZigBee stack profile identifiers. The network discovery will verify whether there is at least one ZigBee router in any discovered network that currently permits joining.

3.4.10.2 Network Formation

The NWK layer, upon receipt of a request from the APL layer, can establish the device as the ZigBee coordinator. The device must be an FFD to act as a ZigBee coordinator. The first step of network formation is performing an ED scan followed by an active scan on a selected number of channels using the MAC management service. The scan request is issued by NLME to MLME. Based on the scan results, the NWK layer picks a frequency channel and a unique PAN identifier. The first channel with the lowest number of existing networks is considered a proper frequency channel to be used in the new network. The NWK layer of the ZigBee coordinator selects 0x0000 as its MAC short address, which is the same as the network address.

The first task of the ZigBee coordinator is configuring the superframe using the MAC management service.

3.4.10.3   Establishing the Device as a Router

A ZigBee router is responsible for routing data frames, route discovery, and route repair. The router can establish its own superframe and accept the requests from other devices to join the network. The request to establish the device as a router is given to the NWK layer by the APL layer using the *NLME-START-ROUTER.request* primitive. Considering that a router can form its superframe, this primitive includes the superframe parameters such as *Beacon Order, Superframe Order*, and *BatteryLifeExtention* (BLE). The NWK layer requests the MAC to create or update the superframe configurations.

3.4.10.4   Joining and Leaving a Network

In any device, if the MAC attribute *macAssociationPermit* is set to TRUE, the device will accept association requests. The NWK layer of a ZigBee coordinator or router can allow other devices to join its network by requesting MLME to set the value of *macAssociationPermit* to TRUE for a fixed period of time. This period is known as the *permit duration*.

The next higher layer can use the *NLME-JOIN.request* primitive to request the NWK layer to join the device to an existing network as either a router or an end device. This primitive also clarifies whether the device was previously associated with this network.

The child chooses the length and the type of the scan (active or passive). The MAC layer of the child delivers the list of discovered networks to the NWK layer. The child then picks a suitable parent. A parent is considered suitable if it allows association and the value of the link cost between the parent and the child is less than 3. If there is more than one suitable parent, the parent in the lowest depth from the ZigBee coordinator must be selected. Upon selection of a suitable parent, the NWK layer of the child initiates the association procedure using the *MLME-ASSOCIATE.request* primitive.

When the parent receives the association request, it will determine whether the device requesting to join is already in the parent network as a child by looking up its neighbor table. If this child is not found in the neighbor table, the child will receive a unique network address. Each parent has only a limited number of addresses available to allocate to its children. If the request to join is granted, the parent will update its neighbor table to add this device as its own child.

If the device was previously associated with this parent, the NWK *rejoin request* command will be used. A child can rejoin its parent even if the parent currently does not accept any new child.

An alternative method to join a network is the *direct join*. The direct join is used when the parent is already preconfigured with the 64-bit addresses of its children. In this case, the child does not attempt to find a suitable parent, because its parent is already selected for it. The parent will initiate the join by searching its neighbor table to identify whether the 64-bit address of the child is already listed in the table. If a match is found in the neighbor table, no further action is required from the parent. If the address is not found in the neighbor table, the parent will create an entry in the table if the table is not full.

Although in direct join the parent does not contact the child, the child is responsible to start an orphan procedure (Section 3.3.4.2.8) to complete the child/parent relationship establishment.

Removing a child from the network can be initiated by the child itself or its parent. The MAC layer disassociation is used to remove a device from the network. When a device leaves a network, all its children can be removed from the network as well. These removed children can join new parents in the network or join another network, depending on the application scenario.

If the device that plans to remove itself from the network is the ZigBee coordinator or a router, the NWK layer *leave* command frame (Section 3.4.11) is broadcast to the entire network by selecting the broadcast address of 0xffff as the destination address. The reason for broadcasting the *leave* command is to let all the devices that count on this specific router or coordinator know that they need to update their routes or find new parents if necessary. A ZigBee end device, in contrast, unicasts the *leave* command only to its parent. In both cases, the APL layer uses *NLME-LEAVE.request* to request the NWK layer initiate the removal procedure.

When the parent decides to remove its child, it unicasts the *leave-request* command to the child. After the child is successfully removed from the network, the parent updates its neighbor table accordingly. The address of the previous child can be reused only if the APL layer allows the address reuse in the *NLME-LEAVE.request* primitive given to the NWK layer for the removal of this child. If the removed child is a ZigBee router, the child will broadcast a *leave* command by selecting the destination address equal to 0xffff.

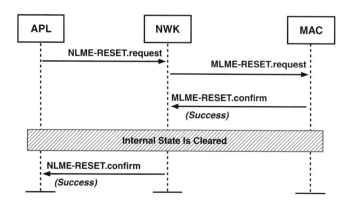

Figure 3.39: The Network Layer Reset Sequence Chart

3.4.10.5 Resetting the NWK Layer

The NWK layer, upon request of the next higher layer, will perform a reset operation (Figure 3.39). The NWK layer first resets the MAC layer. After receiving the MAC reset confirmation, the NWK layer clears all NIB attributes, routing tables, and route discovery tables to their default values. The reset request comes from the APL layer in the form of *NLME-RESET.request.* The NWK layer confirms the result of the reset operation by the *NLME-RESET.confirm* primitive to the APL layer. A device performs the NWK layer reset after the initial powerup, before attempting to join, and after leaving a network.

3.4.10.6 Synchronization

A device can use the synchronization procedure to synchronize or extract pending data from a ZigBee coordinator or router. There are two synchronization scenarios: beacon enabled and nonbeacon enabled. Figure 3.40 shows the sequence charts for both cases. The value of *macAutoRequest* is set to TRUE to ask the MAC to generate and send a data request command automatically. The APL layer uses the *NLME-SYNC.request* to request the NWK layer to initiate the synchronization and data request process. The result of synchronization is delivered to the APL layer by *NLME-SYNC.confirm.*

3.4.11 The NWK Layer Frame Formats

The NWK layer data and command frames and their functionalities are reviewed in this section. We start with the general NWK layer frame format, followed by the data and command frame formats.

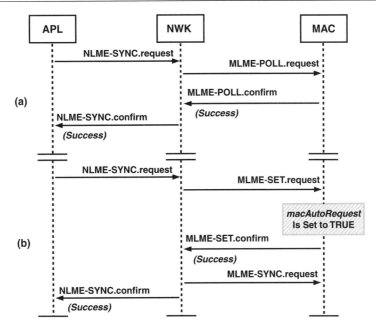

Figure 3.40: Synchronization in (a) A Nonbeacon-enabled Network and (b) A Beacon-enabled Network

The general NWK layer frame is shown in Figure 3.41. The first field is the frame control. The frame type determines if this is a NWK data frame or a NWK command frame. The ZigBee standard evolves over time and the ZigBee protocol version used in a specific device is stored in *nwkcProtocolVersion*. The value of *nwkcProtocolVersion* is always copied into the protocol version subfield of a NWK frame.

The discover route subfield determines the routing option for this frame. If the discover route subfield is set to suppress or enable and a route is already established to the destination, the frame will be sent to the next hop. But if there is no route established to the destination and the discover route subfield is set to suppress, the device will not start a new route discovery. The frame will be discarded or buffered until the route becomes available. If the discover route is set to enable, a route discovery will be initiated if there is no route to the destination. Finally, if the route discovery is set to force route discovery, a route discovery will be initiated for transmitting this frame, even if there is a route already established to the destination. In ZigBee-Pro, the force route discovery option is removed from the discover route sub-field.

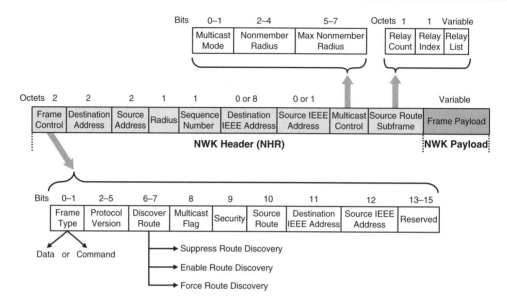

Figure 3.41: General Network Frame Format

If the multicast flag is set to one, the frame will be sent using multicast. The security subfield is set to one if the NWK layer security is enabled. The source route subfield in the frame control is set to one if the source route subframe field is included in the frame. Source routing is a technique in which the sender of a packet can specify the route that a packet should take through the network. The source route subframe contains the list of 16-bit short addresses of the nodes that will be used to relay the frame in source routing. The relay index is set to zero at the source device and is incremented every time the frame is relayed by a router. The relay count is the number of times the frame is relayed.

The NWK layer can optionally include the 64-bit IEEE address in the NWK layer frames if the IEEE address subfields are set to one. The 16-bit network address of the source and destination are always included in the frame. The radius field will determine the maximum number of hops the frame is allowed to have. If the radius parameter is not provided, the radius field of the NWK header is set to twice the value of the *nwkMaxDepth* attribute. The sequence number helps keep track of the sequences transmitted by a device. The value of the sequence number is incremented every time a new frame is transmitted.

The multicast control field only exists if the frame is multicast. The multicast mode subfield determines whether the frame is being sent by the device in nonmember mode (multicast mode equal to 00) or member mode (multicast mode equal to 01). The nonmember radius subfield limits the number of times a multicast frame is rebroadcast by nonmember devices.

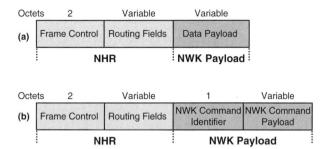

Figure 3.42: The NWK Layer (a) Data Frame Format and (b) Command Frame Format

Table 3.12: NWK Commands

Command Frame Identifier	Command
00000001	Route request
00000002	Route reply
00000003	Route error (network status)
00000004	Leave
00000005	Route record
00000006	Rejoin request
00000007	Rejoin response
00000008	Link status (ZigBee-Pro)
00000009	Network report (ZigBee-Pro)
0000000A	Network update (ZigBee-Pro)

The nonmember radius is decremented every time the frame is rebroadcast by a nonmember device. When the nonmember radius field becomes zero, the frame is no longer rebroadcast by nonmember devices. However, if the content of the nonmember radius is equal to 0x07, there is no limit on the number of times that the frame can be rebroadcast by nonmember devices. Every time a member device (re)broadcasts the frame, it will copy the content of the max nonmember subfield into the nonmember radius subfield.

The data and command frame formats are shown in Figure 3.42. The routing field is the combination of fields between the frame control and the NWK payload in Figure 3.41. The NWK commands are listed in Table 3.12. Each command is identified by an 8-bit number known as the *NWK command identifier.* The NWK command identifier and the command payload will form the NWK frame payload.

The NWK commands are reviewed here briefly. The formats of the NWK layer command payloads in ZigBee 2006 are shown in Figure 3.43. The first command is the *route*

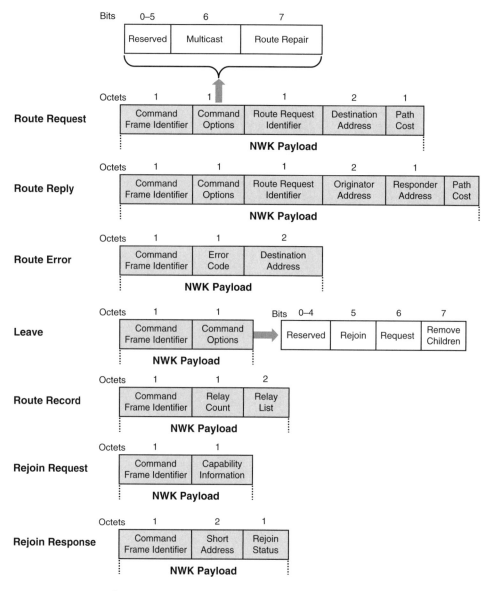

Figure 3.43: NWK Layer Commands Formats

request command, which is used in the route discovery and route repair procedures. If the multicast subfield is set to one, this route request is part of a multicast route discovery. The route repair subfield is set to one to identify this *route request* command as a route repair instead of a route discovery.

The reason for distinguishing route repair from route discovery is that the route discovery is always initiated by the source device. The route repair, in contrast, is initiated by the router device that has tried to relay the message on behalf of the source device but has faced a link failure while attempting to send the frame to the next hop. The route request identifier is an 8-bit sequence number used to identify each route request issued by a source device. The source device increments the content of the route request identifier field every time it generates a new route request. During the route discovery, the routers that broadcast the route request on behalf of the source device keep the route request identifier unchanged. The purpose of the route discovery is to find a route to the destination address provided in the route request command frame.

The *route request* command has a path-cost field to accumulate the path cost from the source device to the destination device. The path-cost field will be used to decide the optimal path from the source to the destination. The destination device may receive the same route request frame from multiple routes. The destination device will send its reply back on the best route to the source device.

In ZigBee-Pro, there is one more field after the path-cost field in the route request command: the destination IEEE address. Also, in ZigBee-Pro bits 3–4 in the command options filed are used to support many-to-one route requests. If the value of bits 2–3 is equal to 0, it means that the route requests is not a many-to-one route request. If the value is equal to 1, the route request is many-to-one and the sender supports a route record table. Finally, if the value of bits 2–3 in the command option field is 2, the route request is many-to-one and the sender does not support a route record table.

The next NWK layer command is the *route reply* command, which is sent by the destination device to the source device in response to a *route request* command. The command option field is similar to the command option field of the *route request* command. The route request identifier is the same as the *route request* command frame. In this way, when the source device receives the *route reply* command back from the destination device, the source device will know this *route reply* command is given in response to one of the source device route requests. The originator address field contains the 16-bit address of the device that originated the route request (source device). The responder address field contains the 16-bit NWK address provided in the *route request* command destination address. The source device always attempts to establish a route to the responder device. The path cost in the *route reply* command is used to accumulate the total path cost from the responder device (destination device) back to the initiator device (source device). In ZigBee-Pro, there are two additional fields after the path-cost field: originator IEEE address and responder IEEE address.

The *route error* command is used to let the source device know about an error in relaying the frame toward a specific destination address. The error code provided in the error code field can be used to determine the cause of the routing error. Examples of the reasons for routing errors are tree-link failure, lack of routing capacity, and low battery level. In the latter, the relaying device is not capable of acting as a router, because its remaining battery energy is critically low. In ZigBee-Pro, this command is renamed to *Network Status* Command and the error code is referred to as the status code. The network status command in ZigBee-Pro contains all the error codes in route error command plus additional codes to report incidents such as PAN identifier update.

A device will transmit the *leave* command either when the device itself intends to leave the network or if the device is requesting another device to leave the network. If the device itself is leaving the network, the request subfield of the *leave* command is set to zero. Otherwise the request subfield is set to one to indicate that the sender of this command is asking the command recipient to leave the network. The device may leave its parent but rejoin another device in the network. If the device intends to rejoin the network, the rejoin subfield of the *leave* command is set to one. Finally, if the device that leaves the network is a parent and the remove children subfield of the command is set to one, all the children of this parent will be removed from the network when their parent leaves.

The *route record* command is transmitted to record the address of all the devices that relay this command frame to the destination along an established route. The relay count subfield value is equal to the number of times the frame was relayed. The short address of all the devices that relayed this frame is kept in the relay list. When a device other than the destination device receives this command, it will add its 16-bit short address to the relay list and increment the relay count before sending it to the next hop via unicast. The address of the next hop is provided in the routing table.

If a device loses its connection to its network—for example, when a child can no longer communicate with its parent—it can use the *rejoin request* command to rejoin the network through a device other than its original parent. In the *rejoin request* command, the device provides the list of its capabilities. This is similar to the list of capabilities provided in a MAC layer association request (Figure 3.25). The device that receives the rejoin request will reply using the *rejoin reply* command. If the new parent has capacity to accept a new child, the short address field in the *rejoin reply* command will contain the new 16-bit address assigned to the child. The rejoin status field is similar to the association status field in Figure 3.25.

The NWK layer in ZigBee-Pro has three additional commands that are not available in ZigBee-2006. The first one is the *Link Status* Command. Each router can use the link status command to communicate its link-cost to other neighboring routers. The second command is the *Network Report* Command, which is used to report network events such as PAN ID conflict and channel condition to a designated device in the *nwkManagerAddr*. The last command is the *Network Update* Command. This command is used by a designated device (identified by *nwkManagerAddr* attribute) to broadcast configuration changes. The format of these commands is provided in the ZigBee specification [3].

3.4.12   Summary of the NWK Layer Responsibilities

The NWK layer is responsible for the following operations:

- Configuring a new device. For example, a new device can be configured to begin its operation as a ZigBee coordinator or try to join an existing network.

- Starting a new network.

- Joining and leaving a network.

- Applying NWK layer security.

- Routing frames to their destination. Only ZigBee coordinators and routers can relay messages.

- Discovering and maintaining routes between devices. This is the ability to discover and record paths through the network for efficient routing of the messages.

- Discovering one-hop neighbors. These are the devices that can be reached directly without using any other device relay service.

- Storing pertinent one-hop neighbor information.

- Assigning addresses to devices joining the network. Only ZigBee coordinators and routers can assign addresses.

3.5 The APL Layer

The application (APL) layer is the highest protocol layer in a ZigBee wireless network. The ZigBee APL layer consists of three sections, shown in Figure 3.44: the application support (APS) sublayer, ZigBee Device Objects (ZDO), and the application framework.

The application support sublayer (APS) provides an interface between the network layer (NWK) and the application layer (APL). The APS sublayer, similar to all lower layers, supports two types of services: data and management. The APS data service is provided by APS Data Entity (APSDE) and is accessed through the APSDE Service Access Point (SAP). The management capabilities are offered by APS Management Entity (APSME) and are accessed through APSME-SAP.

The APS sublayer constants and attributes start with *apsc* and *aps*, respectively. The APS attributes are contained in the APS Information Base (APS IB or AIB). The list of APS constants and attributes is provided in the ZigBee specification [3].

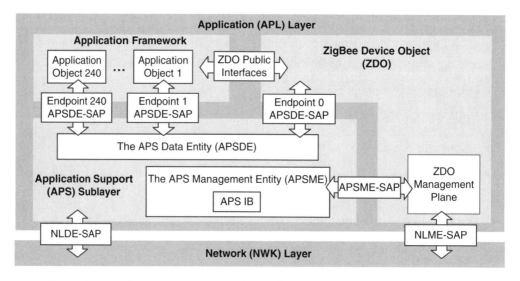

Figure 3.44: The APL Layer Consists of the APS Sublayer, ZDO, and the Application Framework

The application framework in ZigBee is the environment in which application objects are hosted to control and manage the protocol layers in a ZigBee device. Application objects are developed by manufacturers, and that is where a device is customized for various applications. There can be up to 240 application objects in a single device.

The application objects use APSDE-SAP to send and receive data between peer application objects (Figure 3.44). Each application object has a unique endpoint address (endpoint 1 to endpoint 240). The endpoint address of zero is used for the ZDO. To broadcast a message to all application objects, the endpoint address is set to 255. Endpoint addressing allows multiple devices to share the same radio. In the light control example in Section 2.1.4, multiple lights were connected to a single radio. Each light has a unique endpoint address and can be turned on and off independently.

The ZigBee Device Objects (ZDO) provide an interface between the APS sublayer and the application framework. The ZDO contains the functionalities that are common in all applications operating on a ZigBee protocol stack. For example, it is the responsibility of the ZDO to configure the device in one of three possible logical types of ZigBee coordinator, ZigBee router, or ZigBee end device. The ZDO uses primitives to perform its duties and accesses the APS sublayer Management Entity via APSME-SAP. The application framework interacts with the ZDO through the ZDO public interface.

The details of application framework, ZDO, and APS sublayer are reviewed in the following three subsections.

3.5.1   The Application Framework

The ZigBee standard offers the option to use *application profiles* in developing an application. The use of an application profile allows further interoperability between the products developed by different vendors for a specific application. For instance, in a light control scenario, if two vendors use the same application profile to develop their products, the switches from one vendor will be able to turn on and turn off the lights manufactured by the other vendor. The application profiles are also referred to as *ZigBee profiles.*

Each application profile is identified by a 16-bit value known as a *profile identifier*. Only the ZigBee alliance can issue profile identifiers. A vendor that has developed a profile can request a profile identifier from the ZigBee alliance. The ZigBee alliance evaluates the proposed application profile and if it meets the alliance guidelines, a profile identifier will be issued. The application profiles are named after their corresponding application use. For example, the home automation application profile provides a common platform for vendors developing ZigBee-based products for home automation use.

The general structure of an application profile is shown in Figure 3.45. The application profile consists of two main components: clusters and device descriptions. A *cluster* is a set of attributes grouped together. Each cluster is identified by a unique 16-bit number called a *cluster identifier*. Each attribute in a cluster is also identified by a unique 16-bit number known as a *attribute identifier*. These attributes are used to store data or state values. For example, in a temperature control application, a device that acts as the temperature sensor can store the value of the current temperature in an attribute. Then another device that acts as the furnace controller can receive the value of this attribute and turn on or turn off the furnace accordingly. The application profile does not contain the cluster itself. Instead, the application profile has a list of the cluster identifiers. Each cluster identifier uniquely points to the cluster itself.

The other part of an application profile is the *device descriptions* (Figure 3.45). The descriptions provide information regarding the device itself. For example, the supported frequency bands of operation, the logical type of the device (coordinator, router, or end device), and the remaining energy of the battery are provided by the device descriptions. Each device description is identified by a 16-bit value. The ZigBee application profile uses the concept of *descriptor data structure*. In this method, instead of including the

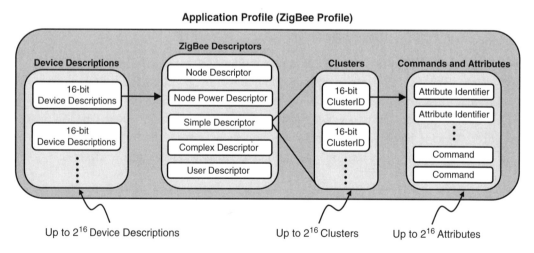

Figure 3.45: The Application Profile

data in the application profile, a 16-bit value is kept and acts as a pointer to the location of the data. This pointer is referred to as the *data descriptor*. When a device discovers the presence of another device in the network, the device descriptions are transferred to provide the essential information regarding the new device.

The device descriptions consist of five sections: node descriptor, node power descriptor, simple descriptor, complex descriptor, and user descriptor. The *node descriptor* provides information such as the node logical type and the manufacturer code. The *node power descriptor* determines whether the device is battery powered and provides the current level of the battery. The profile identifier and clusters are provided in the *simple descriptor*. The *complex descriptor* is an optional part of the device descriptions and contains information such as the serial number and the device model name. Any additional information regarding the device can be included as the *user descriptor*. The user descriptor can be up to 16 ASCII characters. For example, in a light control application, the user descriptor field of a wall switch installed in a hallway can read *Hall switch*.

The node descriptor fields for ZigBee-2006 are provided in Figure 3.46. The node descriptor is a mandatory part of the device descriptions. The logical type can be ZigBee coordinator, router, or end device. The complex descriptor and user descriptor are optional and if their corresponding fields in the node descriptor are set to zero, they are not provided as part of the device descriptions. The APS flag field determines the APS sublayer capabilities. The frequency band (868 MHz, 915 MHz, or 2.4 GHz) is specified in the frequency band field. The MAC capacity flags field is the same as the MAC capacity field presented before in Figure 3.25. A manufacturer can request and receive a manufacturer code from the ZigBee alliance. This code is included in the node descriptor. The maximum size of the APS Sublayer Data Unit (ASDU), in octets, is specified in the maximum buffer size field. The maximum size of a single message that can be transferred to or from a node is provided in the maximum transfer size field (in octets). In ZigBee-Pro, the maximum incoming transfer size and maximum outgoing transfer size are two separate fields (16 bits each).

The server mask field provides information regarding the system server capabilities of this node. A *server* is a device that provides specific services to other devices in the network. If each bit is set to one, the device has the corresponding capability shown in Figure 3.46. The *trust center* is the device trusted by devices within a network to distribute security keys for the purpose of network and end-to-end application configuration management. The security features are reviewed in Section 3.6.

Field Name	Length (Bits)
Logical type	3
Complex descriptor available	1
User descriptor available	1
Reserved	3
APS flag	3
Frequency band	5
MAC capacity flags	8
Manufacturer code	16
Maximum buffer size	8
Maximum transfer size	16
Server mask	16

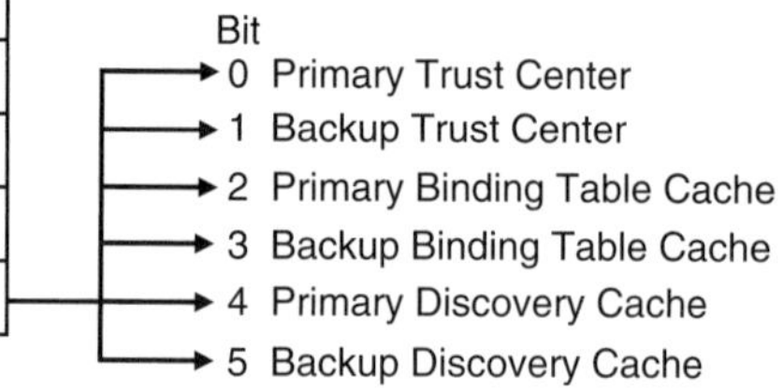

Bit
0 Primary Trust Center
1 Backup Trust Center
2 Primary Binding Table Cache
3 Backup Binding Table Cache
4 Primary Discovery Cache
5 Backup Discovery Cache

Figure 3.46: Node Descriptor Fields

The *primary binding table cache* is a device that allows other devices to store their binding tables with it as long as it has storage space left. The binding procedure is further clarified in this subsection. The primary binding table cache can be used to back up the content of binding tables and restore them whenever necessary. A device can choose to keep its own binding table, known as a *source binding table*, instead of storing it with a primary binding table cache. However, any device can store a backup of the source binding table in the primary biding table cache device and recover it later if necessary.

A ZigBee network may have a *primary discovery* cache device. This device is a ZigBee coordinator or router used to store the descriptors such as node descriptors and power descriptors of some other devices. An end device, for example, that sleeps for long durations can store its descriptors in the primary discovery cache device. If a device in the network tries to locate the information regarding this sleeping end device while the device is inactive, it can get the information from the primary discovery cache device instead. If a network contains sleeping ZigBee end devices, the network must have at least one primary discovery cache device.

The node power descriptor fields are shown in Figure 3.47. The receiver can stay in ON mode while the device is in idle. This might not be a proper option for battery-powered devices. Alternatively, the device can turn off the receiver and only turn it on periodically. The other option is to turn the receiver on by an external trigger whenever there is a need

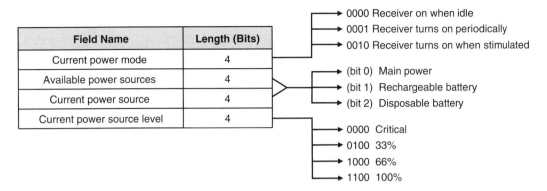

Field Name	Length (Bits)
Current power mode	4
Available power sources	4
Current power source	4
Current power source level	4

0000 Receiver on when idle
0001 Receiver turns on periodically
0010 Receiver turns on when stimulated

(bit 0) Main power
(bit 1) Rechargeable battery
(bit 2) Disposable battery

0000 Critical
0100 33%
1000 66%
1100 100%

Figure 3.47: Node Power Descriptor Fields

to receive a message. The node may have multiple power sources available. For each available power source, the corresponding bit in the available power sources field is set to one. For example, if the node has both main power and rechargeable battery, the available power sources field will read *0011*. The current source of power is specified in the current power source field.

Table 3.13 is the list of simple descriptor fields. The endpoint field contains the endpoint address of a device within the node. The application profile identifier that is supported by this endpoint is in the application profile identifier field. The device description supported by this device is specified by a 16-bit value provided in the application device identifier field. The device description itself may change over time, and the application device version field determines which version of the device description is supported by this device. All the cluster identifiers supported by the device are included in the simple descriptor.

Figure 3.48 illustrates the use of cluster identifiers (*clusterIDs*) in building binding relationships. *Binding* is the task of creating logical links between applications that are related. Devices logically related in a binding table are called *bound devices*. In this example, two wall switches share the same radio. Therefore both switches share an IEEE address and network address. The switches are distinguished by different endpoint addresses. Each switch can have its own application object. Each application object can be accessed independently through its corresponding endpoint address. These two switches control three separate lights. All three lights are also connected to a single radio and each light has a unique endpoint address. A cluster can be an *input cluster* or an

Table 3.13: Simple Descriptor

Field Name	Length (Bits)
Endpoint	8
Application profile identifier	16
Application device identifier	16
Application device version	4
Reserved	4
Application input cluster count	8
Application input cluster list	$16 \times i$ (i = input cluster count)
Application output cluster count	8
Application output cluster list	$16 \times o$ (o = output cluster count)

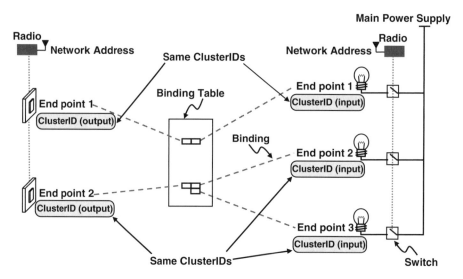

Figure 3.48: The Binding Relationships in a Light Control Example

output cluster. In the binding process, two devices are matched if both devices have the exact same *clusterIDs* but one is an input cluster and the other one is an output cluster. The wall switch at endpoint 1 and the lamp at endpoint 1 have the exact same *clusterIDs* and therefore are considered bound devices. The wall switch at endpoint 2 is bounded to both lamps at endpoint 2 and endpoint 3. The information regarding these logical links is stored in a binding table.

The bind itself can be created by the installer. For example, the installer can push two physical buttons: one on the wall switch and one on the light itself to create the bind between these two devices. This would create an entry in the binding table corresponding to these two devices. More specifically, pushing the button on the wall switch initiates the transmission of the *end device bind request* command (*End_Device_Bind_req*) to the ZigBee coordinator. This command is part of the ZigBee device profile discussed in the next section. After receiving this command, the ZigBee coordinator waits for a period of time to receive the *end device bind request* command from the light. If the second bind request arrives before the timeout period, the ZigBee coordinator matches these two devices based on their profile identifier and the list of their input/output clusters. This is known as a *simple binding mechanism*. In simple binding, user intervention is used to identify the device pairs.

3.5.2   The ZigBee Device Objects

Figure 3.49 shows the ZigBee Device Objects (ZDO) as an interface between the APS sublayer and the application framework. The ZDO are responsible for initializing the APS, NWK, and Security Service Provider (SSP). Similar to the application profiles defined in the application framework, there is a profile defined for the ZDO, which is known as the *ZigBee Device Profile* (ZDP), or simply the *device profile*. The device profile contains device descriptions and clusters, but the device profile clusters do not employ attributes. The ZDO itself has configuration attributes, but these attributes are not included in the device profile. Another difference between the device profile and any application profile is that the application profile is created for a specific application, whereas the device profile defines capabilities supported by *all* ZigBee devices. The device profile has only one device description. The clusters are divided into two groups of mandatory and optional clusters. The mandatory clusters must be implemented in any ZigBee device.

The device profile provides support for device and service discovery as well as binding management. *Device discovery* is the ability to determine the identity of other devices on the PAN. In *service discovery*, the device requests that another device in the network provide detailed information such as its profile identifier or its ZigBee descriptors (e.g., node descriptor or simple descriptor). The device can also request the list of input and output clusters of another device. This cluster list can be used to match devices in the binding procedure.

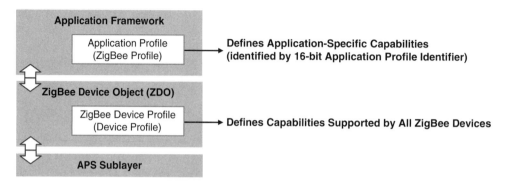

Figure 3.49: The ZDO Acts as an Interface Between the Application Framework and the APS Sublayer

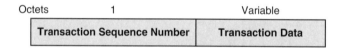

Figure 3.50: The ZigBee Device Profile Command Format

The device profile can be configured as a client and/or a server. A *client* is a device that requests a service such as device discovery or binding. The device that responds to the request and provides the service acts as a *server*. As a result, the services the device profile offers are divided into two categories: *client services* and *server services*. Both of these services are provided in the form of commands with unique cluster identifiers (*clusterIDs*). For example, the *clusterID* of 0x0002 in the device profile is equivalent to the *Node_Desc_req* (*node descriptor request*) command, which is utilized to request the node descriptor of another device. In message exchange between the client and the server, the client is referred to as the *local device* and the server is known as the *remote device*.

The ZigBee Device Profile (ZDP) commands are sent using the APS data service. The format of a ZDP command is shown in Figure 3.50. The first part is an 8-bit transaction number. Any application object maintains a counter and increments it every time a new transaction is transmitted. The content of this counter is copied into the transaction sequence number field of the ZDP command. The transaction data contains the command itself and any data associated with the command. The complete list of ZDP commands and brief descriptions of the commands are provided in six tables in Appendix B.

There are two main groups of commands: client services and server services. The ZDP commands in each group are divided in three categories: device and service discovery, bind management, and network management. The commands in these three categories form three of the objects in the ZDO: the device and service discovery object, the binding manager object, and the network manager object.

The device and service discovery commands allow a device to request information such as NWK address and list of descriptors of any other device in the network. They also allow a device to store its own descriptors in a primary discovery cache device or configure the user descriptor of another device in the network. The bind management commands allow a device to create or remove binding relationships, store binding tables on a primary binding table device, create backup binding tables, and recover previously stored backup binding tables. The network management commands are used to identify nearby networks, request content of routing and neighbor tables, and manage joining and leaving devices in the network.

There are two more objects in the ZDO: the network manager and the security manager. The network manager contains the networking-related primitives to interface with NLME (e.g., *NLME-JOIN.request*). These primitives were reviewed in Section 3.4.10. The security manager object contains security-related primitives to interface with the APS sublayer Management Entity (APSME). The role of the security manager is further clarified in Section 3.6.

The ZDO contain configuration attributes, but these attributes are not related to the ZigBee Device Profile. The ZDO attributes start with :*Config_* and contain information such as node descriptor and the network security level. The complete list of the ZDO configuration attributes is provided in the ZigBee specification document [3].

3.5.3   *The APS Sublayer*

The APS sublayer provides data service to both application objects and ZigBee Device Objects through the APS sublayer Data Entity (APSDE). The APSDE receives the data that needs to be transmitted in the form of a Protocol Data Unit (PDU) from either ZDO or an application object. The APSDE adds proper headers to the PDU to create an APS data frame, which will be passed down the NWK layer.

The APS sublayer Management Entity (APSME) contains primitives to perform three tasks: bind management, APS Information Base (AIB) management, and group management. The binding primitives (*APSME-BIND.request* and *APSME-UNBIND.*

request) allow the next higher layer to request to bind two devices by creating an entry in the local binding table or to unbind two devices by removing the corresponding entry from the local binding table. The *APSME-GET.request* and *APSME-SET. request* primitives allow the next higher layer to read and write an attribute in the APS Information Base. The group management primitives are used to add (or remove) certain endpoints of the node in a group table.

Unicast, broadcast, and multicast are three means of delivering a message that we discussed in Section 3.4. A fourth option supported by the APS sublayer is known as *indirect addressing*. In indirect addressing, a device with limited resources that is bound with other devices in a network can communicate without knowing the address of the desired destination. Indirect transmissions are sent to the ZigBee coordinator by the source device. The ZigBee coordinator looks up the source address, endpoint address, and *clusterID* from its binding table and retransmits the message to each corresponding destination address/endpoint.

Figure 3.51 is the general APS frame format in ZigBee-2006. In ZigBee Pro, there is an optional field after the APS counter field called "Extended Header". There are three types of APS frame: data, command, and acknowledgment. The frame type subfield in Figure 3.51 determines the type of the frame. The delivery mode subfield indicates the transmission options. If the delivery mode is indirect addressing, the indirect address

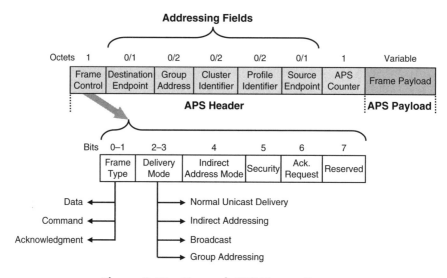

Figure 3.51: General APS Frame Format

mode subfield specifies which address field (source or destination) must be omitted from the frame. If the indirect address mode subfield is set to one, it means that the frame is intended for the ZigBee coordinator, and therefore the destination endpoint field must be omitted. If this field is set to zero, this frame is being sent from the ZigBee coordinator to the destination, and the source endpoint field must be omitted. The security subfield is set by the Security Service Provider (SSP). If the acknowledgment request subfield is set to one, the recipient of the frame must send an acknowledgment back.

If the group address is present, the message will be delivered to all endpoints that are members of the group. The destination endpoint field and the group address field cannot be present concurrently in one frame. The cluster identifier field is only present in a binding operation and contains the cluster identifier that will be used in the binding procedure. The next field contains the intended profile identifier. The APS counter is an 8-bit counter added to each APS frame and incremented every time a new frame is transmitted. This counter helps the frame recipient identify and ignore the duplicate frames.

Figure 3.52 shows the three APS frame types. The data frame has the same format as a general APS frame. The ZigBee specification does not define any APS command other

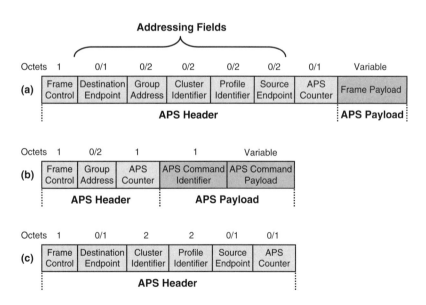

Figure 3.52: APS Layer: (a) Data Frame, (b) Command Frame, and (c) Acknowledgment Frame Formats

than security-related commands, which are covered in Section 3.6. An acknowledgment frame has neither a payload nor a group address. The acknowledgment is never sent to a group of devices.

3.5.4   Summary of the APL Layer Responsibilities

The ZigBee APL layer consists of three parts:

- Application Support (APS) sublayer

- ZigBee Device Objects (ZDO)

- Application framework

The application support (APS) sublayer provides an interface between the network layer (NWK) and the application layer (APL). The following are application support sublayer responsibilities:

- Maintain binding tables.

- Forward messages between bound devices.

- Manage group addresses.

- Map 64 -bit IEEE address to 16-bit network address, and vice versa.

- Support reliable data transport.

ZDO is an application that uses NWK and APS sublayer services to implement a device in one of three ZigBee roles: ZigBee coordinator, router, or device:

- Define the role of the device in the network.

- Discover the devices on the network and their application. Initiate or respond to binding requests.

- Perform security-related tasks.

The application framework in ZigBee is the environment in which application objects are hosted.

3.6   Security Services

In a wireless network, a transmitted message can be received by any nearby ZigBee device. Though in very simple applications the security might not be a big concern, in other applications an intruding device listening to the messages or modifying and resending the messages can cause issues such as violating people's privacy or shutting down operation of a system. In a personnel tracking system, for example, employee privacy can be violated if unauthorized people gain access to employees' exact locations at any given time. To avoid these problems, the ZigBee standard supports the use of encryption and authentication protocols, reviewed in this section.

3.6.1 Encryption

Encryption is the practice of modifying a message by means of substitution and permutation. The ZigBee standard supports the use of Advanced Encryption Standard (AES) [7]. Figure 3.53 shows the basic concept of encryption. The transmitter of the message uses an algorithm to encrypt the message before the transmission and only the intended recipient knows how to recover the original message. The unencrypted message is referred to as *plaintext*. The encrypted message is known as *ciphertext*. If the encryption is done on a block of data, the algorithm is referred to as a *block cipher*. ZigBee uses a 128-bit block cipher. The practice of encrypting and decrypting messages is called *cryptography*.

In AES, each encryption algorithm is associated with a key. The algorithm itself is public knowledge and available to everyone, but the value of the key in each transmission is kept secret. The key is a binary number. There are different methods to acquire a security key. For example, the key can be embedded in the device itself by the manufacturer. Alternatively, a new device that joins a network may get its security key from a designated device in the network. The details of different methods of acquiring security keys are discussed in the remainder of this section.

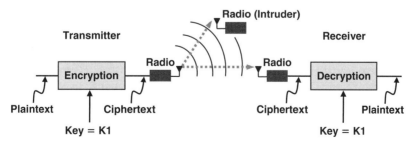

Figure 3.53: Encryption Using Symmetric Keys

The number of bits in a key will determine the level of security. If the key is only 8 bits, the intruder needs to try a maximum of 2^8 ($=256$) different keys to recover the original message. ZigBee supports the use of 128-bit keys, which means there are 2^{128} ($\approx 3.4 \times 10^{38}$) possible keys. It is computationally infeasible to try 2^{128} different keys.

In Figure 3.53, the receiver is using the exact same key as the transmitter to perform decryption, which is known as the *symmetric key* method. The ZigBee standard supports only symmetric key cryptography. The ZigBee standard provides methods to establish keys and share the keys between two or more devices. Considering that the algorithm itself is known by the potential intruder, the main effort is to make sure that the key is not distributed beyond the intended recipients.

Two types of keys are used during secure communication: the link key and the network key. The *link key* is shared between two devices and can be used in unicast communication. The *network key* is shared among the entire network and is employed when a message is broadcast. Any secure ZigBee network has a designated device called the *trust center* that distributes the link keys as well as the network key to other devices. There is only one trust center in each network. The ZigBee coordinator determines the address of the trust center by setting the AIB attribute *apsTrustCenterAddress*.

The AES algorithm itself is a series of well-defined steps that use the provided key to shift and mix a block of data to create an encrypted version of the same block of data. In AES the size of this block of data is always 128 bits. But the key can be 128, 196, or 256 bits. The ZigBee standard uses the 128-bit key option. All these steps are invertible and the receiver that holds the correct key can perform the reverse steps to recover the original message. The AES has been announced as a standard by the National Institute of Standards and Technology (NIST) [7]. According to the committee on national security systems, the design and strength of the AES algorithm is sufficient to protect even classified information up to the SECRET level [8].

There are three methods for a device to acquire a link key: preinstallation, key transport, and key establishment. In the *preinstallation* method, the manufacturer embeds the key in the device itself. For example, DIP switches can be used to select one of the preprogrammed set of keys. In this way, when the device joins a network, it does not need to ask the trust center for the security key. In many applications, this is the most secure way of acquiring a security key. In the *key transport* method, the device asks the trust center for the security key. This request is sent to the trust center using an APS sublayer command (reviewed later in this section). The trust center can send the key to the requesting device over an unsecured link. In this method, there is a moment of vulnerability for the system if this unprotected transmission is received by an intruding device. The solution is to use a *key-transport key*, which is a key used to secure the transmission of any key, other than the master key, from the trust center to the requesting device.

Key establishment is the method of creating a random key in two devices without communicating the key itself over an unprotected wireless link. The key establishment service in ZigBee is based on the Symmetric-Key Key Establishment (SKKE) protocol [3]. The devices that are establishing the key already have a common key known as the *master key*. The master key is provided to the devices by means of preinstallation, key transport, or user-entered data (e.g., passwords). The master key must be installed in each device prior to running the key establishment protocol. The *key-load key* is a dedicated key that can be used to transfer a master key from the trust center to a device in the network. The key establishment is used only to derive a link key and is not used to generate the network key.

In the SKKE protocol there are two devices, an initiator device and a responder device. The *initiator device* establishes a link key using the available master key and transfers a specific data to the responder. The *responder* uses the data and derives the link key. The initiator also derives the link key from the same data. If the derivation is done correctly, the two devices have the same link key that can be used in the symmetric key cryptography. The details of SKKE protocol can be found in the ZigBee specification [3].

Among the main constraints in implementing security features in a ZigBee wireless network are limited resources. The nodes are mainly battery powered and have limited computational power and memory size. ZigBee is targeted for low-cost applications, and the nodes hardware may not be tamper resistant. If an intruder acquires a node from an operating network that has no tamper resistance, the actual key could be obtained simply from the device memory. A tamper-resistant node can erase the sensitive information including the security keys if tampering is detected.

The trust center has two modes of operation: commercial mode and residential mode. In commercial mode, the trust center *must* maintain a list of devices, master keys, link keys, and network keys. All the incoming NWK frames are checked for freshness. A frame is considered fresh if it is not a duplicate frame. The size of memory required for the trust center in commercial mode grows with the number of devices in the network. Residential mode, in contrast, is designed for low-security residential applications. In residential mode the only key that must be maintained in the trust center is the network key. Keeping the list of devices and other security keys is optional. Also, checking for the freshness of all NWK frames is not required. The memory size required for the trust center does not grow with the number of devices in residential mode.

In ZigBee networking, each protocol layer (APS, NWK, and MAC) is responsible for the security of the frames initiated by that layer. Considering that all protocol layers normally reside in the same node, for simplicity the same security key is used by the APS, NWK, and MAC layers in a single node.

3.6.2   Authentication

The ZigBee standard supports both device authentication and data authentication. *Device authentication* is the act of confirming a new device that joins the network as authentic. The new device must be able to receive a network key and set proper attributes within a given time to be considered authenticated. In *data authentication*, the receiver verifies if the data itself has been altered or changed.

3.6.2.1   Device Authentication

The device authentication procedure is performed by the trust center. When a device joins a secure network, it has the status of "joined, but unauthenticated." If the trust center decides not to authenticate the newly joined device, the trust center will request the device be removed from the network.

The authentication procedure is different in residential and commercial modes. In residential mode, if the new device that joins the network does not have a network key, the trust center needs to send the network key over an unprotected link, which causes a moment of vulnerability. If the new device already has the network key, it must wait to receive a dummy (all-zero) network key from the trust center as part of authentication procedure. The new device does not know the address of the trust center and uses the source address of this received message to set the trust center address in its APS Information Base (*apsTrustCenterAddress*). The joining device is now considered authenticated for residential mode.

In commercial mode, in contrast, the trust center never sends the network key to the new device over an unprotected link. But the master key may be sent unsecured in commercial mode if the new device does not have a shared master key with the trust center. After the new device receives the master key, the trust center and the new device start the key establishment protocol. The new device has a limited time (*apsSecurityTimeOutPeriod*) to establish a link key with the trust center. If the new device cannot complete the key establishment before the end of the timeout period, the new device must leave the network and retry the association and authentication procedure again. When the new link key is confirmed, the trust center will send the network key to the new device over a secured connection. The joining device is now considered authenticated for commercial mode.

3.6.2.2 Data Authentication

The purpose of data authentication is to make sure the data is not changed in transit. To achieve this goal, the transmitter accompanies the frame with a specific code known as the Message Integrity Code (MIC). The MIC is generated by a method known to both receiver and transmitter. An unauthorized device will not be able to create this MIC. The receiver of the frame will repeat the same procedure and if the MIC calculated by the receiver matches the MIC provided by the transmitter, the data will be considered authentic. The level of data authenticity is increased by increasing the number of bits in the MIC. The ZigBee and IEEE 802.15.4 standards support 32-bit, 64-bit, and 128-bit MIC options.

Data authenticity is different from data confidentiality. In confidential data transmission, the data is encrypted before transmission and therefore only the intended recipient will be able to recover the data. Confidential data may be accompanied by a Message Integrity Code, but it is not required. The intended recipient of confidential data with no MIC is confident that the data has not been recovered by an intruder, but the data may have been changed in transit. On the other hand, data can be transmitted unencrypted but accompanied with a MIC. In this case, the data is not confidential, but its authenticity can be verified.

The MIC is also referred to as Message Authentication Code (MAC) or authentication tag. The ZigBee and IEEE 802.15.4 standard documents use MIC instead of MAC to avoid confusion with the Message Authentication Code (MAC) and the Medium Access Control (MAC). The MIC in ZigBee is generated using the enhanced Counter with Cipher Block Chaining Message Authentication Code (CCM*) protocol. The CCM* is defined to be used in conjunction with 128-bit AES and shares the same security key with AES.

Figure 3.54 shows the role of AES-CCM* in data authentication and confidentiality. On the transmitter side, the plaintext in the form of 128-bit blocks of data enters the AES-CCM*. The responsibility of the AES-CCM* is to encrypt the data and generate an associated MIC, which is sent to the receiver along with the frame. The receiver uses the AES-CCM* to decrypt the data and generate its own MIC from the received frame

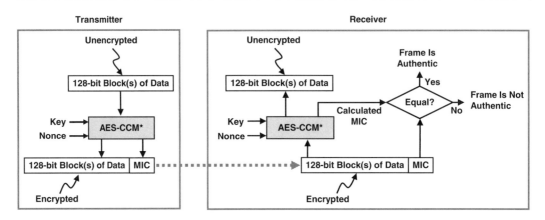

Figure 3.54: Application of the Message Integrity Code (MIC) in Data Authentication

to be compared with the received MIC. The CCM* is referred to as a generic mode of operation that combines the encryption and data authentication. The CCM* offers encryption-only and integrity-only capabilities.

In Figure 3.54, there are three inputs to the AES-CCM*: the data itself, the security key, and the *nonce*. The nonce is a 13-octet string constructed using the security control, the frame counter, and the source address fields of auxiliary header. The auxiliary header fields and the CCM* nonce are shown in Figure 3.55. The AES-CCM* uses the nonce as part of its algorithm. The value of the nonce is never the same for two different messages using the same security key, because the frame counter is incremented every time a new frame is transmitted. The use of the nonce ensures *freshness* of the received frame. The motivation to use the nonce is that an intruder, without the security key, is capable of receiving a secured message and simply resending the exact message after a period of time. This retransmitted message will have all the correct security features of a valid message, but the frame counter will indicate that the frame was received previously. In this way, the frame counter helps identify and prevent processing of duplicate frames. This is referred to as *checking the frame freshness*. If the intruding device changes the frame counter associated with the frame before retransmitting the frame, the receiver device will notice this unauthorized modification when it compares the calculated and received MICs (Figure 3.54).

The Cyclic Redundancy Check (CRC), discussed in Section 3.3.5.1.1, helps identify any error in a received frame. The CRC is designed to detect only the accidental error

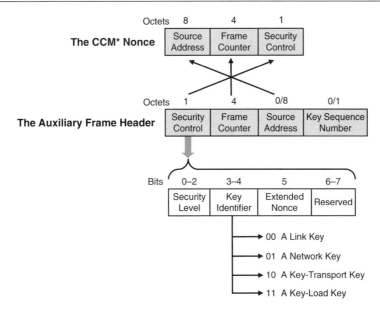

Figure 3.55: The Auxiliary Header Format and the CCM* Nonce

in the received data, and the CRC can be reproduced by any intruder. A MIC provides stronger assurance of authenticity compared to the CRC. The MIC generated by the CCM* detects intentional and unauthorized modifications of the data as well as accidental errors.

3.6.3   The Auxiliary Frame Header Format

The security features of a frame are identified in an optional auxiliary header that is added to an outgoing frame. Figure 3.55 illustrates the format of the auxiliary frame header. The security-level subfield defines eight unique security options listed in Table 3.14. If the security level is set to 000, the data is not encrypted and no MIC will accompany the outgoing frame. All security levels other than 000 and 100 include MIC for data authentication. The size of the MIC can be 32 bits, 64 bits, or 128 bits. The value of *M* in Table 3.14 refers to the size of the MIC in octets. The data encryption has only one option for the size of the key (128 bits). If the security attribute contains *ENC*, it means that the data is *encrypted*. For simplicity, all the devices in a network will have the same level of security. If a particular device requires a higher level of security, it must create its own network with the proper security level.

Table 3.14: The Security Levels Available for MAC, NWK, and APS

Security Level	Security Attributes	Data Encryption	Frame Integrity
000	None	OFF	NO (M=0)
001	MIC-32	OFF	YES (M=4)
010	MIC-64	OFF	YES (M=8)
011	MIC-128	OFF	YES (M=16)
100	ENC	ON	NO (M=0)
101	ENC-MIC-32	ON	YES (M=4)
110	ENC-MIC-64	ON	YES (M=8)
111	ENC-MIC-128	ON	YES (M=16)

The key identifier subfield in Figure 3.54 specifies the type of key used to protect the frame. The source address field is an optional field and is only included if the extended nonce subfield is set to one. The frame counter indicates the freshness of the frame and is incremented every time a new frame is transmitted. The key sequence number is only present when the network key is used to secure the frame. The combination of the source address, frame counter, and security control fields of the auxiliary header form the CCM* nonce, which is an input used by the CCM* protocol for authenticated encryption operation.

Figure 3.56 shows that the frame security can be added at the APS, NWK, and MAC layers. The security settings of each layer are stored as attributes in their corresponding Information Bases (IB). At each layer, the CCM* provides an encrypted payload and an encrypted MIC. The MIC provides an authenticity check for the payload, the auxiliary header (AUX HDR), and the header of the layer in which the security is implemented. For example, if the frame is secured only at the APS sublayer, the APS header (APS HDR) is used as an input to generate the MIC, but the NWK header and MAC header are not included. The security materials such as keys, frame counts, and security level are stored in an access control list (ACL). The ACL is used to prevent unauthorized devices from participating in the network. The ACL is stored in MAC PAN Information Base (PIB) and is accessed and modified similar to other MAC attributes.

3.6.4   The APS Sublayer Security Commands

Table 3.15 lists the APS security commands in ZigBee-2006. The frame formats of these commands are shown in Figure 3.57. The first four commands are used in the

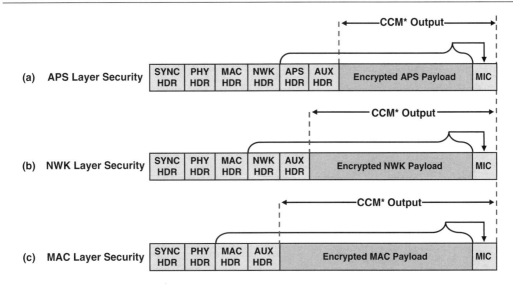

Figure 3.56: The Frame Security can be Added at the APS, NWK, and MAC Layers

Table 3.15: The APS Security Commands

Command Identifier	Value
APS_CMD_SKKE_1	0x01
APS_CMD_SKKE_2	0x02
APS_CMD_SKKE_3	0x03
APS_CMD_SKKE_4	0x04
APS_CMD_TRANSPORT_KEY	0x05
APS_CMD_UPDATE_DEVICE	0x06
APS_CMD_REMOVE_DEVICE	0x07
APS_CMD_REQUEST_KEY	0x08
APS_CMD_SWITCH_KEY	0x09

SKKE protocol. The initiator and responder addresses are the addresses of two devices that participate in a key establishment. The SKKE protocol is reviewed in the ZigBee specification [3]. The transport-key command carries the actual key to a device in the network. There are four types of keys that can be carried by this command. The *trust*

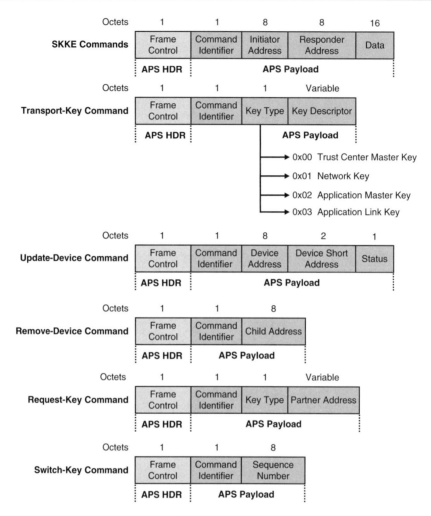

Figure 3.57: The APS Security Commands

center master key is the master key used to establish a link key between the trust center and any other device in the network. The *key descriptor* associated with this key includes the master key itself, the destination address, and the source address. The *network key* key descriptor includes the destination and source addresses as well as a sequence number. The *application master key* is used to set up a link key between any two devices. The key descriptor for the application master key and the link key contains the address of the recipient of the key and a flag that is set to one if the recipient of this frame has requested the key.

The *update device* command changes the status of a device in a network. There are three status options: the device has joint security, the device has joint insecurity, or the device has left the network. The *remove device* command requests to remove a child from the network. The *request key* command is used to request a specific key. The key type field is similar to the key type field of the transport key command. The partner address field is only present if an application link key is being transmitted. Whenever an application link key is requested, the trust center will send the key to the device that initiated the key request as well as the address provided in the partner address field. The *switch key* command is used to activate a network key in a device. The network keys are identified by a sequence number.

In ZigBee-Pro, there are 5 more commands in addition to the commands shown in Table 3.15. Four of these additional commands are used for Entity Authentication, which allows two devices to mutually authenticate each other. The fifth additional command allows a device to send a command to a device that lacks the current network key. The details are provided in ZigBee specification document [3].

3.6.5 Security Attack Examples

This section reviews two examples of security vulnerabilities applicable to AES-CCM* protocol: the same-nonce attack and the denial-of-service attack.

When a nonce is used as part of AES-CCM*, encrypting the same plaintext two times will result in two different ciphertexts because the nonce will be different even if the same key is reused. This property is known as *semantic security*. If, for any reason, the ACL provides the same nonce and the same security key for two consecutive messages, an eavesdropper will be able to recover partial information regarding the plaintexts. This is known as the *same-nonce attack*. For example, if the nonce and the key used in generating two cipher texts are the same, the exclusive-OR (XOR) of these two ciphertexts will be the same as the XOR of their corresponding plaintexts [9].

One of the occasions on which a same-nonce attack can happen is after a power failure that results in a clear ACL [9]. If the last nonce states are unknown after the power failure, the system might reset the nonce states to a default value. This reset action increases the chance of reusing the same nonce with a key that has been used before the power failure and could lead to the same-nonce attack vulnerability. Storing the nonce states in a nonvolatile memory (NVM) and recovering them after each power failure is one way of protecting the system from same-nonce attacks due to power failure.

A *denial-of-service (DoS) attack* causes a node to reject all received messages [9]. A system is vulnerable to a DoS attack if the message integrity is not verified, even if the communicated messages are encrypted. The attacker composes a message that includes random content as the encrypted payload (without knowing the security key) and sets the frame counter to the maximum. This message is sent by the intruder to a recipient device. The message receiver will decrypt the payload to a random plaintext, which might not have any meaning to the next upper layer. The main harm in this attack is setting the frame counter to the maximum. The received frame counter is always used to set a high-water mark for the frame counter associated with the authorized originator of the frame. By setting the high-water mark of the frame counter to the maximum, any legitimate frame that arrives after the DoS attack will be automatically rejected by the recipient device because the frame counter of the received message will be less than the high-water mark established during the DoS attack, assuming that the received frame is a duplicate. Validating message authenticity helps prevent a DoS attack.

3.6.6   Summary of the Security Services

The ZigBee standard supports the use of the following optional security services:

- Encryption for data confidentiality

- Device and data authentication

- Replay (duplicate frame) protection

Five different keys can be used in a secure network:

- The *link key* is shared between only two devices and can be used in unicast communications.

- The *network key* is shared among the entire network and can be utilized when a message is broadcast.

- The *master key* is used to establish a link key between two devices as part of key establishment protocol.

- The *key-transport key* is a key used to secure the transmission of any key, other than the master key, from the trust center to the requesting device.

- The *key-load key* is a key used to secure the transmission of the master keys.

References

[1] Open Systems Interconnection Basic Reference Model: The Basic Model ISO/IEC 7498-1:1994.

[2] IEEE std 802.15.4-2006: Wireless Medium Access Control (MAC) and Physical Layer (PHY) Specifications for Low-Rate Wireless Personal Area Network (WPANS), Sept. 2006.

[3] ZigBee Specification 053474r17, Jan. 2008; available from www.zigbee.org.

[4] F. Tobagi and L. Kleinrock, "Packet Switching in Radio Channels: Part II, The Hidden Terminal Problem in Carrier Sense Multiple-Access and the Busy-Tone Solution," IEEE Transaction on Communication, Dec. 1975, pp. 1417–1433.

[5] International Telecommunication Union; available from www.itu.int.

[6] A. Tanenbaum, "*Computer Networks*," 4th ed., Prentice Hall, 2002.

[7] Advanced Encryption Standard (AES), Federal Information Processing Standards Publication 197, U.S. Department of Commerce/N.I.S.T, Springfield, VA, Nov. 26, 2001; available from http://csrc.nist.gov/.

[8] CNSS Policy No. 15, Fact Sheet No. 1, June 2003; available from www.cnss.gov.

[9] N. Sastry and D. Wagner, "Security Considerations for IEEE 802.15.4 Networks," ACM WiSe 2004, Oct. 2004, pp. 32–42.

Transceiver Requirements

This chapter presents the knowledge essential to understanding and comparing the performance of different transceivers. The modulations and spreading methods available in IEEE 802.15.4 are also reviewed. In a wireless sensor network (WSN), the sensor output is an analog signal and needs to be converted to digital. The basics of analog–to-digital converters (ADCs) and their performance metrics are provided. A transceiver requires a crystal oscillator to generate a precise clock; therefore, crystal oscillator operation and analysis are covered in this chapter.

4.1   Typical IEEE 802.15.4 Transceiver Building Blocks

A *transceiver* provides the means of radio communication for two nodes. Examples of the blocks that can be found along with the radio interface include microcontrollers; memory; analog-to-digital converters, or ADCs (for sensor interface); and general-purpose input/ output ports (GPIOs). Figure 4.1 shows a block diagram of a typical transceiver. The function of each block is briefly described in this section.

An *antenna* converts the electromagnetic waves to electric currents, and vice versa. The type of antenna depends on the application and is discussed in Chapter 5 in detail. The majority of antennas have 50 Ohm single-ended impedances. Matching and filtering are the passive components that can improve receiver performance during receive mode (for instance, enhancing the receiver sensitivity) and during transmit mode can reduce the out-of-band emission of the output signal. The transceiver does not receive and transmit simultaneously, so the antenna can be shared between the receiver and transmitter paths by means of a transmit/ receive (T/R) switch. A *crystal* is used as part of the precise clock generation circuitry and can be inside or outside the IC package. Crystal selection considerations, including safety margin and basic operation of crystal oscillators, are discussed in Section 4.11.

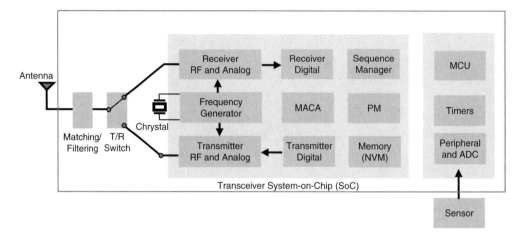

Figure 4.1: Block Diagram of a Typical 802.15.4 Transceiver

In the receiver path, the radio frequency (RF) and analog blocks amplify the received signal, down-convert the signal to a lower frequency, and filter the unwanted frequency components. The receiver digital block demodulates the signal and extracts the binary information, which is passed to the PHY protocol layer. The frequency generator is responsible for providing a high-accuracy reference clock and a low-accuracy (low-power) clock for power-saving modes of operation. This block also generates a high-frequency signal that the receiver path will use to down-convert the signal and by the transmitter to generate the output signal.

In the transmitter path, the digital section of the transmitter converts the packets into modulated signals. The RF portion of the transmitter up-converts, amplifies, and filters these modulated signals to the desired programmable output power and ensures that the transmitted signal complies to the local emission regulations. The *sequence manager* (SM) is responsible for managing the priority of events, coordinating the timing, and keeping track of the state of the IC at all times. The universal asynchronous receiver/transmitter (UART), serial peripheral interface (SPI), and GPIOs are examples of typically available peripherals in a transceiver IC. Most sensors have analog outputs. An on-chip ADC simplifies sensor information gathering.

Considering that these transceivers are built for low-current, low-duty-cycle applications, the power management unit is fairly simple and normally consists of a number of voltage regulators and may include a simple buck converter as well. Separate regulators for RF and digital units will help reduce the chance of the digital noise leaking to sensitive RF blocks. An introduction to buck converters is provided in Chapter 6.

The *MAC accelerator* (MACA) is hardware that provides low-level MAC and PHY link controls and may include buffers for transmit and receive packets. It not necessary for a transceiver to have a MACA unit, but a MACA can reduce the CPU load and allow some of the functions to be executed independently of the processor. Timers are necessary to provide real-time interrupt (RTI). Normally there is more than one timer in a transceiver.

Nonvolatile memory (NVM), random access memory (RAM), and read-only memory (ROM) are examples of memory options for a System-on-Chip (SoC) transceiver. NVM retains the stored information even after it is disconnected from the power source. The content of NVM is rewriteable. RAM, in contrast, loses its information when the power source is switched off. ROM is a low-cost, one-time programmable memory that keeps its information when it is not connected to a power source.

4.2 Receiver Sensitivity

In the IEEE 802.15.4 standard, receiver sensitivity is defined as the lowest received signal power that yields a packet error rate (PER) of less than 1% [2]. IEEE 802.15.4 requires only $-85\,$dBm of sensitivity for operations in the 2.4 GHz ISM band. In the 868/915 MHz band, if the BPSK modulation is used, the required sensitivity is $-92\,$dBm. The optional modes of operation in the 868/915 MHz band (using ASK and OQPSK modulation) must meet $-85\,$dBm of sensitivity.

If a 50 Ohm single-ended antenna is used, the sensitivity level of $-85\,$dBm translates to a signal with effective (or rms) voltage of 12.6 uV:

$$\text{Signal power} = -85\,\text{dBm} = 10^{\left(\frac{-85-30}{10}\right)} = 3.16 \times 10^{-12} = 3.16\,\text{pW}$$

$$\text{Signal voltage} = \sqrt{P \times R} = \sqrt{3.16 \times 10^{-12} \times 50} = 12.6\,\text{uV}$$

This means that any received signal with root-mean-square (rms) voltage of 12.6 uV or higher can be detected and the data can be extracted with a PER of less than 1%. During this test, no interference (Section 4.3) is present. Also, the PHY Service Data Unit (PSDU) length of the data packets used in this test must be 20 octets. Although it's not required by the standard, the available commercial transceivers are capable of delivering $-95\,$dBm to $-100\,$dBm of sensitivity level.

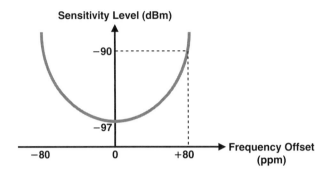

Figure 4.2: Example of Receiver Sensitivity Degradation Due to Frequency Offset

Based on IEEE 802.15.4, receiver sensitivity is measured with a signal that has 0% error vector magnitude (EVM), whereas in reality, the signal may have an EVM of up to 35%, which is the maximum EVM allowed in IEEE 802.15.4. The EVM, discussed in Section 4.7, is an indication of the modulation inaccuracy. The EVM can worsen the sensitivity level, but as long as the EVM is below 35%, the sensitivity degradation should be less than 2dB.

In IEEE 802.15.4, the carrier frequency offset of the input signal is zero during the receiver sensitivity test [2]. In reality, however, the input signal has some frequency offset and if the receiver does not use a frequency offset correction mechanism, the sensitivity level will be degraded. Figure 4.2 shows an example of receiver sensitivity degradation versus input carrier frequency offset when there is no offset correction.

IEEE 802.15.4 not only defines the receiver minimum input signal lever (receiver sensitivity), it also determines the receiver maximum input level. A receiver must be able to receive signals with a power level of -20 dBm or higher and maintain PER of less than 1% to be compliant with IEEE 802.1.5.4.

4.3   Adjacent and Alternate Channel-Jamming Resistance Tests

As the number of IEEE 802.15.4-enabled devices installed in residential and commercial places increases, there is a possibility of two or more nodes transmitting signals at the same time. Consider the example in Figure 4.3, where node B is using channel 20 (2450 MHz) to transmit packets to node A. At the same time, the adjacent channel 21 (2455 MHz) is occupied with the transmitted signals from node C, which are not intended

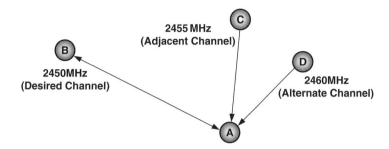

Figure 4.3: Example of Communication in the Presence of Adjacent and Alternate Interferers. Each Circle Represents an IEEE 802.15.4-Enabled Transceiver

for node A. It is expected from the receiver in node A to have some level of tolerance for an occupied adjacent channel and still be able to detect the desired signal and yield a PER of 1% or better.

In the 2.4 GHz band, the channels at ± 5 MHz of the desired channel are called the *adjacent channels*. The channels at ± 10 MHz of the desired channel are known as the *alternate channels*. In our previous example, channels 19 and 21 are adjacent channels. Channels 18 and 22 are alternate channels for the desired channel 20.

The performance of a receiver in the presence of adjacent and/or alternate channels is evaluated using the *jamming resistance* tests. In the IEEE 802.15.4 adjacent channel jamming test (see Figure 4.4a), both the desired signal and the adjacent channel interferer have equal power (-82 dBm), and the desired signal is a pseudorandom IEEE 802.15.4-compliant modulated signal. The receiver must be able to detect and demodulate the desired signal with PER of 1% or less to be compliant with IEEE 802.15.4 [2].

Figure 4.4b is the alternate channel-jamming resistance test required by the standard. The alternate channel power is 30 dB higher than the desired signal power, and a standard compliant receiver must be able to have PER of 1% or less when receiving the desired and alternate channel signals concurrently. These jamming resistance tests are also known as *adjacent and alternate channel rejection* tests. The preceding discussion applies to both 2450 MHz and 915 MHz bands with the same level of adjacent and alternate channel rejection requirements, but it does not apply to the 868 MHz band, because there is only one channel available in the 868.0–868.6 MHz band. In the 915 MHz band, the adjacent channels and alternate channels are ± 2 MHz and ± 4 MHz away from the desired channel.

The commercially available transceivers can perform much better than the minimum required by the standard. It is common for a transceiver to have a PER of less than 1%

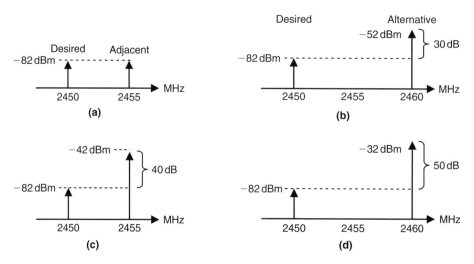

Figure 4.4: The Input Signals for the Adjacent and Alternate Jamming Tests: (a) and (b) Are IEEE 802.15.4 Standard Requirements; (c) and (d) Are Jamming Resistance Levels of Typical Commercial Receivers

when an adjacent or an alternate channel with signal power of 40 dB to 50 dB higher than the desired signal is present. Even though in the standard jamming resistance tests there is only one interferer at a time, it is wise to check the performance of the receiver when both adjacent and alternate channels are present.

4.4   The Modulation and Spreading Methods for 2.4 GHz Operation

Any time domain signal has an equivalent representation in the frequency domain. The graph of the signal power versus frequency is referred to as the *signal power spectral density* (PSD). Figure 4.5a represents a typical IEEE 802.15.4 signal PSD centered at 2450 MHz. This center frequency is known as the *carrier frequency*. The signal bandwidth is normally considered the frequency band that contains the majority of the signal power. There are different definitions of the signal bandwidth in the literature. In Figure 4.5a, the power spectral density peak is at the middle of the graph. The 3 dB corners of the PSD graph are the frequencies at which the PSD value drops 3 dB below its maximum. The 3 dB bandwidth is defined as the frequency interval between the 3 dB corners. Another way to define the signal bandwidth is to consider the nulls on the PSD graph as the signal borders. This is known as the *null-to-null bandwidth* of the signal. The

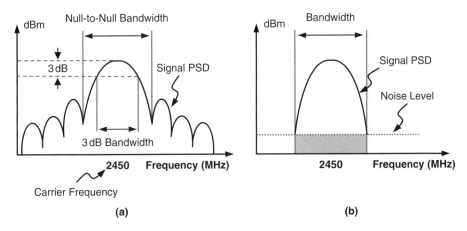

Figure 4.5: (a) Signal Bandwidth Definitions and (b) Signal-to-Noise Ratio (SNR) Definition

frequency band that contains 99% of signal power is referred to as the *99% bandwidth* of the signal.

When the destination receives the signal, the receiver circuit will use a combination of analog and digital filtering to remove the spectral contents outside the frequency band of interest. The noise is normally modeled as a signal with flat power spectral density (Figure 4.5b). When the signal is filtered, the only noise that matters is the noise within the frequency band of interest. The ratio of the total signal power to total noise power within the band of interest is called the *signal-to-noise ratio* (SNR). The SNR is an indication of the signal quality; increasing SNR will improve receiver PER if the receiver is not suffering from multipath issues.

Regardless of the definition of the signal bandwidth, the bit rate associated with a signal is proportional to the signal bandwidth in the frequency domain. That means that for a given modulation/spreading method and signal SNR, to double the bit rate we need to double the signal bandwidth. The *spectral efficiency* is defined as bits/second/Hertz and determines the relationship between the signal bit rate and the associated bandwidth for a given modulation and spreading technique. For example, if a method provides 2 bits/second/Hertz spectral efficiency, it means that a signal with 1 MHz bandwidth can deliver a bit rate of up to 2 Mbps.

IEEE 802.15.4 uses spreading methods to improve the receiver sensitivity level, increase the jamming resistance, and reduce the effect of the multipath. (The multipath issue is discussed in the next chapter.) The spreading method required by IEEE 802.15.4 for the

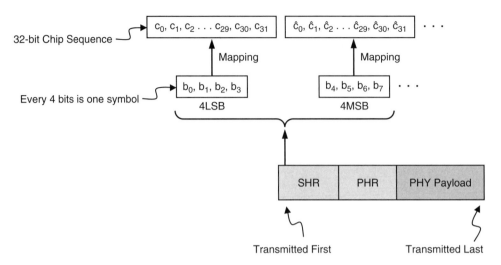

Figure 4.6: The Signal Spreading in DSSS

2.4 GHz frequency band is the Direct Sequence Spread Spectrum (DSSS). In an IEEE 802.15.4-specific implementation of DSSS, every 4 bits of each octet of a PHY Protocol Data Unit (PPDU) are grouped together and referred to as a *symbol* (see Figure 4.6). Then a lookup table is used to map each symbol to a unique 32-bit sequence. This 32-bit sequence is also known as the *chip sequence* or the *pseudorandom noise (PN) sequence*. The table of these symbol-to-chip mappings is provided in Appendix C.

Since every 4 bits are mapped to a unique chip sequence, the lookup table contains 16 chip sequences. At first glance, each chip sequence appears as a random sequence of zeroes and ones (i.e., random noise). But each sequence is selected by a procedure to minimize its similarity to the other 15 sequences. The similarity of two sequences is measured by calculating the *cross-correlation function* of two sequences. The cross-correlation is determined by multiplying the sequences together and then calculating the summation of the result. A sequence containing 0 and 1 is replaced by its bipolar (-1 and $+1$) before calculation of cross-correlation. If $x(n)$ and $y(n)$ are two sequences, the cross-correlation of these two sequences is the following:

$$r_{xy}(0) = \sum_{n=-\infty}^{n=+\infty} x(n)y(n) \qquad (4.1)$$

The $r_{xy}(0)$ is the calculated cross-correlation of $x(n)$ and $y(n)$ when neither of the sequences is shifted. The higher the absolute value of $r_{xy}(0)$, the higher the similarity of two sequences.

If the cross-correlation $r_{xy}(0)$ is equal to zero, it indicates that the sequences $x(n)$ and $y(n)$ are as dissimilar as possible. In this case, the sequences $x(n)$ and $y(n)$ are known as *orthogonal sequences*. The 16 sequences used in IEEE 802.15.4 are not completely orthogonal and are referred to as *near-orthogonal* or *quasi-orthogonal sequences*.

The cross-correlation can be calculated for $x(n)$ and $y(n–k)$ as well, where $y(n–k)$ is the sequence $y(n)$ shifted by k:

$$r_{xy}(k) = \sum_{n=-\infty}^{n=+\infty} x(n)y(n-k) \tag{4.2}$$

In this case, $r_{xy}(k)$ is an indication of the similarity of the $x(n)$ sequence and the $y(n–k)$ sequence.

Since each 4 bits of the actual data are mapped to a 32-bit chip sequence, the effective over-the-air bit rate is increased by a factor of eight. Remembering the fact that the bit rate is proportional to the signal bandwidth, we know that the signal bandwidth will also increase by a factor of eight. The basic concept of signal spreading is shown in Figure 4.7a. If the original signal before spreading has a bandwidth of 250 KHz, after the spreading the bandwidth will be increased to 2 MHz. Since the signal spreading does not add any physical energy to the signal, the peak of the PSD of the signal will be less after the spreading because the same signal energy is distributed over a larger bandwidth.

While the signal with 2 MHz bandwidth is traveling over the air, undesired noise and interferers can be added to it. The receiver of the signal will use the *despreading* procedure to recover the original signal. In despreading, every 32 bits of the received signal are compared to 16 possible chip sequences, and the chip sequence that has the most similarity to the received signal will be chosen as the received chip sequence. From the same lookup table used by the transmitter, the receiver can recover the original 4 bits of the packet.

As a result of the despreading, the signal energy will be concentrated back to the original bandwidth of 250 KHz, but the despreading will not affect the noise level in the 250 KHz band of interest. Since the signal energy is increased in the desired band without increasing the noise level, the effective SNR is increased by signal despreading. This increase in the signal SNR will directly improve the receiver sensitivity. This improvement in SNR is referred to as the *processing gain*. The value of the processing gain is equal to the ratio of the signal bit rate after spreading to the signal bit rate before spreading. For example, the processing gain for the 2.4 GHz mode of operation in IEEE 802.15.4 is equal to 9 dB:

$$\text{Processing Gain} = 10 \times \log_{10}\left(\frac{2 \text{ Mbps}}{250 \text{ Kbps}}\right) \cong 9 \text{ dB} \qquad (4.3)$$

The signal spreading by the transmitter and despreading by the receiver also help reduce the effect of the interferers. In Figure 4.7b, an interferer signal has the same carrier frequency as the desired signal. When both the desired signal and the interferer arrive at the receiver, the despreading will concentrate the desired signal energy back to 250 KHz. However, the despreading procedure results in spreading the interferer signal energy over a larger bandwidth. After a filter removes the spectral contents outside the frequency band of interest (250 KHz), the remaining portion of the interferer inside the frequency

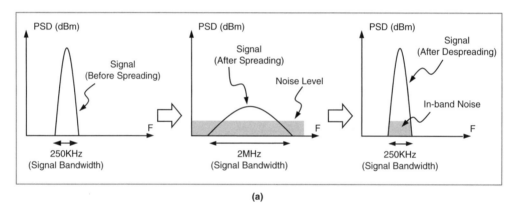

(a)

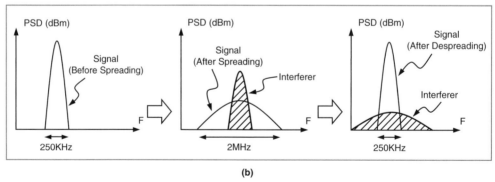

(b)

Figure 4.7: Signal Spreading and Despreading Can Help Improve SNR and Reduce the Effect of Interferers

band of interest has much lower power compared to the original interferer signal. In this way, signal spreading can improve the resistance of a receiver to some interferers. One way to quantify the quality of a signal in the presence of an interferer is the *signal-to-interference ratio* (SIR), which is the ratio of the power of the wanted signal to the total residue power of the unwanted signals in the frequency band of interest:

$$\text{SIR} = \frac{\text{Desired Signal Power}}{\text{Total Interferences Power}} \text{ (dB)} \qquad (4.4)$$

IEEE 802.15.4 requires the use of offset-quadrature phase shift keying (OQPSK) modulation for the 2.4 GHz mode of operation. In a phase shift keying (PSK) modulation, the signal phase is used as a way to transport binary information between the transmitter and the receiver. Each signal phase corresponds to a binary number. The transmitter sets the signal phase before transmission and receiver can recover the binary information by detecting the signal phase. Consider a signal $s(t)$ with carrier frequency of f_C and phase of $\theta(t)$:

$$s(t) = A \times \sin(2\pi f_C t + \theta(t))$$

The amplitude A does not carry any information in PSK. The simplest PSK modulation is BPSK (binary PSK). In BPSK, the phase $\theta(t)$ can only take two discrete values, zero and 180 degrees. These two phases represent binary levels of zero and one. The quadrature PSK (QPSK) allows for four phase options: 45°, 135°, –45°, and –135°. Therefore, each phase can represent 2 bits. The four phases in QPSK correspond to binary numbers 00, 01, 10, and 11. Figure 4.8 shows the relationship between the signal phases and their associated binary numbers in BPSK and QPSK.

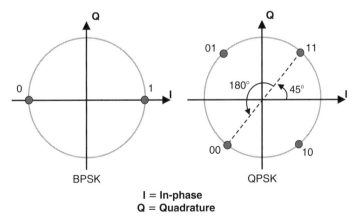

Figure 4.8: The BPSK and QPSK Constellations

One observation in QPSK modulation is the following:

$$
\begin{cases}
\theta = \dfrac{\pi}{4} & \rightarrow s(t) = A \times \sin(2\pi f_C t + \dfrac{\pi}{4}) & = \dfrac{A}{\sqrt{2}} \times (\sin(2\pi f_C t) + \cos(2\pi f_C t)) \\[2mm]
\theta = \dfrac{3\pi}{4} & \rightarrow s(t) = A \times \sin(2\pi f_C t + \dfrac{3\pi}{4}) & = \dfrac{A}{\sqrt{2}} \times (-\sin(2\pi f_C t) + \cos(2\pi f_C t)) \\[2mm]
\theta = \dfrac{-3\pi}{4} & \rightarrow s(t) = A \times \sin(2\pi f_C t + \dfrac{-3\pi}{4}) & = \dfrac{A}{\sqrt{2}} \times (-\sin(2\pi f_C t) - \cos(2\pi f_C t)) \\[2mm]
\theta = \dfrac{-\pi}{4} & \rightarrow s(t) = A \times \sin(2\pi f_C t + \dfrac{-\pi}{4}) & = \dfrac{A}{\sqrt{2}} \times (\sin(2\pi f_C t) - \cos(2\pi f_C t))
\end{cases}
\tag{4.5}
$$

This means that all four possible phase options in QPSK can be built by summation or subtraction of $\sin(2\pi f_C t)$ and $\cos(2\pi f_C t)$ functions. The scaling factor of $\sqrt{2}$ does not change the information embedded in the signals, because the amplitude A does not carry any information. This property is used in implementation of the QPSK modulator in the transmitter. Figure 4.9 shows the basic design of a QPSK modulator in a transmitter. The train of pulses with -1 and $+1$ values are multiplied by $\sin(2\pi f_C t)$ and $\cos(2\pi f_C t)$ functions, and the results are added together to form the modulated signal $s(t)$. The signal $s(t)$ will be amplified by a power amplifier (PA) before transmission. The modulated signal $s(t)$ consists of two components: an in-phase component ($\pm\cos(2\pi f_C t)$) and a quadrature component ($\pm\sin(2\pi f_C t)$).

There are two issues with the simple implementation suggested by Equation 4.5. The first one relates to the shape of the PSD of the signal generated by this QPSK modulator. The frequency spectrum is shared by many users, and it is important to minimize the signal bandwidth and suppress the signal power outside the frequency band of interest. The QPSK modulator in Figure 4.9 requires filtering to shape the signal PSD. These filters are shown between the pulse trains and the multiplier in Figure 4.9. IEEE 802.15.4 requires the use of specific filters (half-sin filter, raised cosine filter, and square-root raised cosine filter). These filters are also referred to as *pulse-shaping filters* because they modify the shape of the binary pulses in Figure 4.9. Figure 4.10 shows the effect of the half-sin pulse-shaping filter. The half-sin filter replaces each sharp pulse with one half of a sinusoid signal:

$$
p(t) = \begin{cases}
\left| \sin\left(\pi \dfrac{t}{2T_C} \right) \right| & 0 \le t \le 2T_C \\[3mm]
0 & \text{otherwise}
\end{cases}
\tag{4.6}
$$

where $2T_C$ is the width of the pulse.

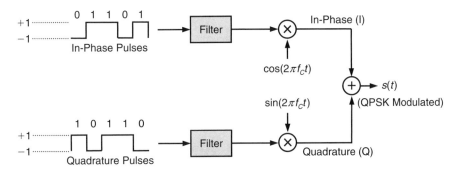

Figure 4.9: The Simplified Diagram of the QPSK Modulator

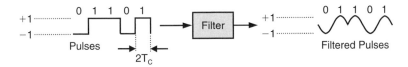

Figure 4.10: The Effect of the Half-Sin Pulse-Shaping Filter

The second problem with the QPSK modulator suggested by Equation 4.5 is the maximum phase shift of $s(t)$ every time a new pair of bits enters the modulator. For example, in Figure 4.8, if the first pair of bits that enters the modulator concurrently is 11 and the next pair is 00, the phase of the modulated signal $s(t)$ will change 180°. This abrupt phase change causes large amplitude variation in $s(t)$. The smaller the amplitude variations, the easier the implementation of the PA. Therefore, the QPSK modulation itself is slightly modified to an Offset-QPSK (OQPSK) to limit the maximum abrupt phase shift in $s(t)$. In Figure 4.9, there is no time delay or offset between the in-phase and quadrature pulses that enter the modulator. In an OQPSK, there is a time offset equal to one-half of the pulse period between the in-phase and quadrature pulses. This small difference between OQPSK and QPSK will limit the maximum instantaneous phase shift of a signal modulated using OQPSK to 90 degrees. This property makes OQPSK a superior choice over QPSK for most practical implementations.

Figure 4.11 represents the operations performed in a transmitter for the 2.4 GHz mode of operation in IEEE 802.15.4. The individual bits are grouped together to form a symbol. Then the symbols are modulated, amplified, and transmitted.

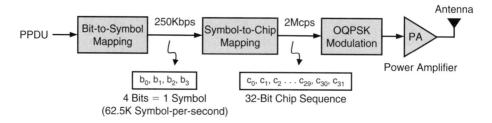

Figure 4.11: The Spreading and Modulation for the 2.4 GHz Operation

Table 4.1: Modulation and Spreading Options in IEEE 802.15.4

	Frequency (MHz)	Modulation	Chip Rate (Kchip/s)	Bit Rate (Kb/s)	Processing Gain (dB)	Spreading Method
Case 1	2400–2483.5	O-QPSK	2000	250	9	16-array DSSS
Case 2	868–868.6	BPSK	300	20	11.76	Binary DSSS
Case 3	902–928	BPSK	600	40	11.76	Binary DSSS
Case 4	868–868.6	ASK	400	250	Measured	20-bit PSSS
Case 5	902–928	ASK	1600	250	Measured	5-bit PSSS
Case 6	868–868.6	O-QPSK	400	100	6	16-array DSSS
Case 7	902–928	O-QPSK	1000	250	6	16-array DSSS

4.5   Modulation and Spreading Methods for 868/915 MHz Operation

In the 868/915 MHz mode of operation in IEEE 802.15.4, there is one mandatory modulation technique and two optional modulation methods. The BPSK is the mandatory modulation. The OQPSK and amplitude shift keying (ASK) are optional. Table 4.1 summarizes these modes of operation. Cases 1, 2, and 3 are the mandatory modes of operation for each frequency band. The optional modes of operation are shaded. Case 1 was discussed in the previous section. If a transceiver operates in the 868/915 MHz band and its goal is to support either of the optional ASK or OQPSK modulation schemes, it must be capable of receiving and transmitting BPSK signals as well.

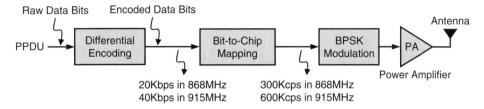

Figure 4.12: The Spreading and Modulation Steps for the 868/915 MHz DSSS/BPSK Mode of Operation

Figure 4.12 shows the modulation and spreading steps for the 868/915 MHz DSSS/BPSK mode of operation (cases 2 and 3). The first step is *differential encoding*, a simple procedure that is used to remove any possible phase ambiguity between the transmitter of a BPSK modulated signal and the signal recipient. Using the absolute value of the signal phase can cause uncertainty because if the phase of the BPSK modulated signal is changed 180 degrees, the signal still appears as a valid BPSK signal. To remove this ambiguity, instead of using the absolute value of the signal phase to recover corresponding information, the phase difference between the received signal and the next arriving signal is used to determine the received information. This is achieved by performing a simple exclusive *OR* operation in the differential encoder. If the data entering the differential encoder is the sequence $R(n)$ and the output of the encoder is $E(n)$, the $E(n)$ is generated from $R(n)$ using the following equation:

$$E(n) = R(n) \oplus E(n-1) \tag{4.7}$$

where $\oplus$ indicates the exclusive *OR* operation. The value of $E(0)$ is equal to zero. For example, if the raw data entering the differential encoder is the following:

$$R = [1\ 0\ 1\ 1\ 0\ 1\ 1]$$

then the encoded output will become:

$$E = [1\ 1\ 0\ 1\ 1\ 0\ 1]$$

When the receiver of the packet receives the encoded data $E(n)$, it can simply recover the original data $R(n)$ by performing another exclusive *OR* operation:

$$R(n) = E(N) \oplus E(n-1) = [1\ 0\ 1\ 1\ 0\ 1\ 1] \tag{4.8}$$

Equation 4.8 indicates that the original data $R(n)$ is recovered by using two consecutive bits of the received encoded information. In other words, the original bit is determined by comparing the phase difference of two consecutive signals instead of using the absolute phase value, which is the main goal of using differential encoding.

In Figure 4.12, the encoded data bits are mapped using the DSSS mechanism. But in contrast to case 1, the bits are not grouped together to form a multibit symbol. Instead, each single bit is mapped to a 15-bit chip sequence. The mapping table is provided in Appendix C. The processing gain associated with cases 2 and 3 is 11.76 dB:

$$\text{Processing Gain} = 10 \times \log_{10}\left(\frac{300\ \text{Kbps}}{20\ \text{Kbps}}\right) = 10 \times \log_{10}\left(\frac{600\ \text{Kbps}}{40\ \text{Kbps}}\right)$$
$$= 10 \times \log_{10}(15) = 11.76\ \text{dB} \tag{4.9}$$

Implementation of BPSK modulation, similar to OQPSK, requires pulse-shaping filters to shape the PSD of the modulated signal.

ASK is used in optional cases 4 and 5. In ASK modulation, the information is embedded in the signal amplitude instead of the signal phase. The spreading method for cases 4 and 5 is parallel sequence spread spectrum (PSSS) instead of the direct sequence spread spectrum (DSSS) used in all other cases.

We can understand the basic concept of PSSS by comparing PSSS to DSSS. Both of these spreading techniques rely on nearly orthogonal sequences to spread the signal before transmission. In DSSS, a single chip sequence is transmitted, but PSSS sends superposition of multiple orthogonal sequences in parallel. Figure 4.13 shows the PSSS mapping mechanism. First, each bit of the binary data that needs to be transmitted is converted to bipolar data. In bipolar data, the logic level 0 is replaced by -1, whereas logic level $+1$ is kept the same. Therefore, the data that enters the multiplier stage is an array with $+1$ and -1 levels. Each bipolar data is multiplied by a unique sequence. The PSSS sequences are identified by sequence numbers. (The table of PSSS sequences is provided in Appendix A.) In the 868 MHz frequency band, the sequence numbers 0 to 19 are utilized. Therefore, the value of parameter n in Figure 4.13 is equal to 20. The 915 MHz band uses only sequence numbers 0 to 4, and the value of parameter n in Figure 4.13 is set to 5. In other words, every 20 bits in 868 MHz are grouped together to form a single symbol. In 915 MHz, every 5 bits form a single symbol.

The result of the summation (i.e., superposition) in Figure 4.13 is a single sequence. Each entry in the summation sequence is a multilevel value. The orthogonal characteristic

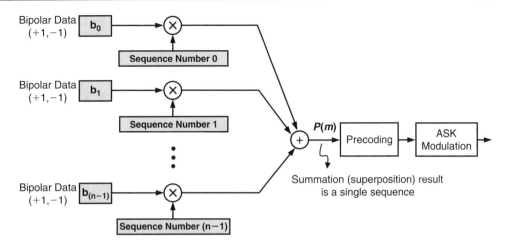

Figure 4.13: The PSSS Modulator in a Transmitter

of the sequences allows the receiver to be capable of recovering each bipolar data from this single sequence. Because the sequences used in PSSS are nearly orthogonal, the PSSS method is also referred to as the *orthogonal code division multiplexing* (OCDM) method. Each near-orthogonal sequence provided for the 868 MHz band in Appendix A has an offset of one-half of a chip compared to the next sequence. Therefore, these 32-bit sequences are sometimes referred to as 64 *half-chip sequences*.

Figure 4.14a shows the beginning of an example of the summation sequence $P(m)$. These different levels will determine the amplitude levels in the ASK modulator. The maximum and minimum of the levels of the sequence $P(m)$ are random numbers. If the maximum and minimum of $P(m)$ are symmetric about zero, the implementation of ASK will be optimal. Therefore, after the summation in Figure 4.13, there is a precoding block. The role of this precoding block is to add a constant value to the sequence $P(m)$ to ensure that the maximum and minimum are symmetric about zero. For example, in Figure 4.14a, where the maximum is $+5$ and the minimum is -3, subtracting a constant 1 from the sequence $P(m)$ will make the maximum and minimum symmetric about zero:

$$\text{adjustment constant} = \frac{\text{Max} + \text{Min}}{2} = \frac{+5 - 3}{2} = 1 \qquad (4.10)$$

The adjusted sequence is shown in Figure 4.14b. The precoding block also normalizes the sequence $P(m)$ to make the maximum and minimum of the sequence $P(m)$ equal to $+1$ and -1, respectively.

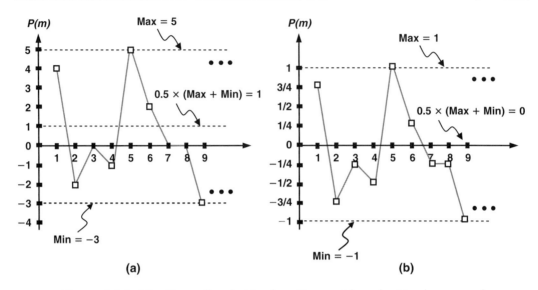

Figure 4.14: The Precoding is Used to Ensure That the Maximum and Minimum are Symmetric About Zero

Figure 4.15 shows that in cases 4 and 5, the PHY header (PHR) and the PHY payload are modulated using PSSS/ASK. But the sequence header (SHR) is directly modulated using the BPSK without any mapping stage. The reason is that the content of SHR is already a spread sequence and does not need any additional spreading. Recall from Chapter 3 that in the PSSS/ASK mode of operation, the preamble is constructed by repeating the sequence number 0 in Appendix A. In 868 MHz, the sequence number 0 is repeated twice to form the preamble sequence. In 915 MHz, the preamble is generated by repeating sequence number 0 six times. The start-of-frame delimiter (SFD) is also generated using the sequences in Appendix A. In both 868 MHz and 915 MHz, the SFD contains the inverted sequence number 0 in Appendix A. Since the content of SHR has only two levels ($+1$ and -1), the ASK modulation is reduced to BPSK modulation. The pulse shaping performed for SHR is the same as the pulse shaping used for PHR and the PHY payload.

Availability of PSSS/ASK modulation as an optional mode of operation in sub-GHz provides the opportunity of increasing the bit rate to 250 Kbps from 40 Kbps (maximum) in DSSS/BPSK mode. In PSSS, in addition to processing gain due to increase in signal bandwidth, there is coding gain as well. The coding gain is the amount of additional SNR that would be required to provide the same BER performance for an uncoded signal. However, comparing the PSSS/ASK versus DSSS/BPSK and DSSS/OQPSK in

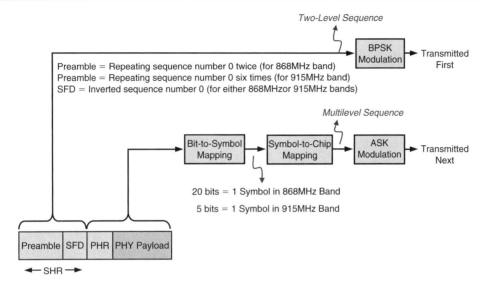

Figure 4.15: In the PSSS Optional Mode of Operation, the SHR is Modulated using BPSK

terms of processing gain will depend on the transceiver nonlinearity performance. The reason is that in the BPSK and OQPSK, in contrast to multilevel ASK modulation, there is no information in the amplitude, and the nonlinearity of the transmitter and receiver have less impact on the quality of the signal. Therefore, the actual SNR improvement for the PSSS/ASK mode of operation must be determined by measurement or simulation.

Cases 6 and 7 in Table 4.1 are based on the DSSS/OQPSK modulation method similar to case 1. But the chip sequences used in cases 6 and 7 are half the size of the length of the chip sequences used in case 1. Every 4 bits (a symbol) in cases 6 and 7 are mapped to a 16-bit chip sequence. (The mapping table is provided in Appendix C.) Since the length of the chip sequence in cases 6 and 7 is less than in case 1, the processing gain is reduced to 6 dB.

4.6 Transmitter Output Power

The transmitted signal needs to satisfy three types of requirements. First, the PSD of the signal needs to meet IEEE 802.15.4 criteria. Second, the signal power needs to be adjustable based on the distance between the nodes. Finally, the transmitted signal needs to comply with local radio spectrum regulations.

4.6.1   Power Spectral Density Limits

IEEE 802.15.4 requires that the output signal PSD meet the criteria summarized in Table 4.2. Figure 4.16 visualizes the PSD limit for operation in the 2.4 GHz band. During PSD measurement, the resolution bandwidth must be 100 KHz. The peak power is the highest average power measured within ± 1 MHz of the carrier frequency in the 2.4 GHz band and within ± 600 KHz of the career frequency in the 915 MHz band. For the 868 MHz band, since there is no adjacent channel, the only criterion is that the signal is filtered by a raised cosine filter before transmission.

4.6.2   Transmit Power Adjustment

IEEE 802.15.4 requires that the transmitter be capable of delivering at least -3 dBm of output power [2]. Reducing the output power will reduce the current consumption during

Table 4.2: IEEE 802.15.4 PSD Limits

Frequency Band	Frequency Offset	Relative Limit	Absolute Limit
2.4 GHz	$\left|f - f_c\right| > 3.5\,\text{MHz}$	-20 dBc	-30 dBm
915 MHz	$\left|f - f_c\right| > 1.2\,\text{MHz}$	-20 dBc	-20 dBm
868 MHz	N/A	N/A	N/A

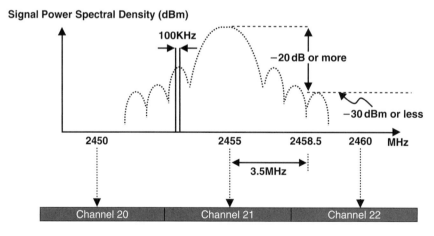

Figure 4.16: PSD Limits in IEEE 802.15.4 (2.4 GHz)

transmit mode and extend the battery life. Even if battery life is not a concern, reducing the transmitter output power to the minimum necessary for a reliable communication link will help reduce the interference to other wireless nodes. A transmitter should have at least 30 dB of output power adjustment range.

4.7 Error Vector Magnitude

The *error vector magnitude* (EVM) is an indication of the modulation accuracy. The discussions in this section apply to all three kinds of modulations possible in IEEE 802.15.4.

In an ideal transmitter, the output signal has all the constellation points at the ideal locations (see Figure 4.17). But in reality, the location of the actual signal on the I-Q diagram may be drifted from the ideal position. We always expect a receiver to tolerate some level of modulation error. To quantify this modulation error, let's start by calculating the amplitude of the error:

$$\text{Error Amplitude} = \sqrt{(I_i - I_A)^2 + (Q_i - Q_A)^2} = \sqrt{\Delta I^2 + \Delta Q^2} \qquad (4.11)$$

I_i and Q_i are the In-phase and Quadrature values of an ideal signal. The actual location of the signal is shown by I_A and Q_A. The root mean square (rms) value of the error amplitude for N symbols is:

$$\text{Error amplitude (rms)} = \sqrt{\frac{1}{N}\left(\sum_{k=1}^{N} \Delta I_k^2 + \Delta Q_k^2\right)} \qquad (4.12)$$

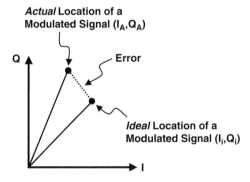

Figure 4.17: Error Vector Magnitude

If the ideal signal amplitude is S, the EVM is defined as:

$$\text{EVM}(\%) = \frac{\text{Error Amplitude (rms)}}{\text{Ideal Signal Amplitude}} \times 100 = \frac{\sqrt{\frac{1}{N}\left(\sum_{k=1}^{N} \Delta I_k^2 + \Delta Q_k^2\right)}}{S} \times 100 \qquad (4.13)$$

IEEE 802.15.4 requires measuring the error amplitude for 1000 chips. The result must show an EVM of less than 35% to pass the EVM requirement of the IEEE 802.15.4 standard [2].

4.8 Symbol Timing

The reference clocks in the receiver and transmitter of two different nodes are not necessarily synchronized when a new wireless link is being established. IEEE 802.15.4 requires the receiver to have a timing recovery mechanism. The receiver uses the synchronization header (SHR) of the received packet to synchronize its clock and lock to the bit stream.

4.9 Frequency Offset Tolerance

IEEE 802.15.4 allows the transmitted signal to have ± 40 ppm of carrier frequency offset [2]. The receiver carrier frequency is also allowed to have up to ± 40 ppm of error. Therefore, the total frequency offset between a receiver and a transmitter can be up to ± 80 ppm. The receiver may implement an automatic frequency control (AFC) unit to compensate for the frequency offset and consequently improve its sensitivity level.

4.10 Turnaround Time

During wireless communication between two nodes, the transceiver might need to switch between transmit (TX) and receive (RX) modes several times. This transition from one mode to another is not instantaneous and can take several microseconds (us). For a transceiver to be IEEE 802.15.4 compliant, it needs to complete any transition between receive and transmit modes in less than 12 symbol periods [2]. For instance, if the transceiver operates in the 2450 MHz band, the symbol rate is 62.5 K symbols/second and each symbol period is 16 μs. Therefore, the transceiver must be able to switch between receive and transmit modes in less than 192 μs (12×16 μs $= 192$ μs). The value of the turnaround time is saved in *aTurnaroundTime*, which is a PHY constant. PHY constants are hardware dependent and cannot be changed during operation.

The exact definition of TX-to-RX turnaround time is the shortest time possible at the air interface from the trailing edge of the last chip (of the last symbol) of a transmitted PPDU to the leading edge of the first chip (of the first symbol) of the next received PPDU. The RX-to-TX turnaround time is defined as the shortest time possible at the air interface from the trailing edge of the last chip (of the last symbol) of a received PPDU to the leading edge of the first chip (of the first symbol) of the next transmitted PPDU [2]. The standard also requires that the TX-to-RX turnaround time be less than or equal to the RX-to-TX turnaround time.

4.11 Crystal Selection Considerations

A *crystal* is used as part of precise clock generation circuitry and can be inside or outside the packaged IC. A crystal is a piezoelectric device that mechanically vibrates whenever a voltage is applied to its two terminals. These mechanical vibrations can be translated into an equivalent electrical circuit (see Figure 4.18). In the motional arm of the crystal, R_m, C_m, and L_m are called *motional parameters*. Capacitor C_0 is called a *shunt capacitor* and models the parasitic capacitance due to the packaging of the crystal.

To discuss topics such as crystal safety margin, we need to know the basics of crystal oscillator operation. The material in this section is not intended to prepare the reader to be a crystal oscillator designer. Instead, it provides the basis for understanding, comparing, and selecting the right crystal for an application.

Figure 4.19a shows the simplified block diagram of a typical crystal oscillator, which consists of an amplifier and a filter. At startup there is no oscillation and the only input to the amplifier is the existing noise in the circuit. The amplifier increases the noise power and delivers the amplified noise to the next stage, which is a very sharp (narrowband) filter around the nominal frequency of the crystal. The output of the filter is then

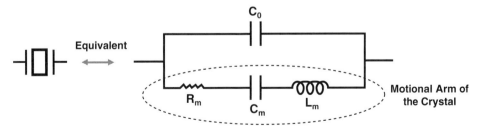

Figure 4.18: Electrical Equivalent of a Crystal

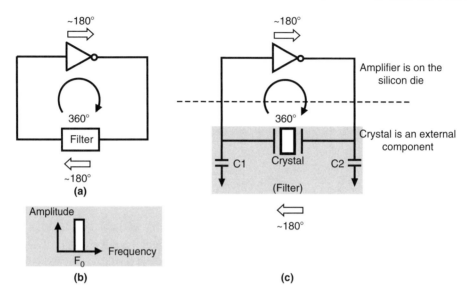

Figure 4.19: (a) A Typical Crystal Oscillator Consists of an Amplifier and a Filter. (b) Filter Frequency Response has a Very Narrow Bandwidth. (c) A Crystal Oscillator with Load Capacitances

amplified again, and this process is repeated until the signal delivered to the input of the filter becomes large enough to force the amplifier into a nonlinear region and therefore reduce the amplifier gain. When the oscillator reaches the steady state, the loop gain (the gain of the amplifier times the gain [loss] of the filter) is equal to one. To ensure a stable oscillation, the total phase shift that the loop provides needs to be N × 360°. Normally the amplifier is implemented as an inverter and shifts the signal 180°. Therefore the filter stage needs to provide another 180° phase shift to bring the total phase shift to 360°. Practically, the amplifier stage phase shift is not exactly 180°; therefore, the filter stage needs to compensate for this phase difference and ensure that the total phase shift of the loop equals 360°.

In a crystal oscillator, the filter stage is implemented using a crystal that can include load capacitances C_1 and C_2 as well. A crystal has two resonant frequencies: the *series resonant frequency* (f_s) and the *antiresonant frequency* (f_A). At the series resonant frequency (f_s), the impedance between the crystal's two terminals is at its minimum. The impedances of C_m and L_m cancel each other out at the series resonant frequency; therefore, the motional arm of the crystal has only resistive impedance (R_m). At the antiresonant frequency (f_A), the inductance of L_m reacts with all capacitors in the crystal and increases the total impedance between the terminals of the crystal to its maximum. The frequency of f_A is only slightly higher than f_s.

The total resistance of the crystal at the series resonant frequency is defined as *effective series resistance* (ESR, or R_{ESR}), which is not necessarily equal to R_m. The value of R_{ESR} is normally given in the crystal datasheet and also can be calculated from the equation:

$$R_{ESR} = R_m \left(1 + \frac{C_0}{C_{Load}} \right)^2 \tag{4.14}$$

where:

$$C_{Load} = \left(\frac{C_1 \times C_2}{C_1 + C_2} \right) \tag{4.15}$$

The load capacitances of C_1 and C_2 include the PCB and contacts parasitic capacitances as well.

4.11.1 Safety Factor

Another way to analyze the crystal oscillator behavior is the negative resistance concept. At the series resonant frequency, the crystal and the loads can be represented by R_{ESR} (Figure 4.20a). This means that the crystal oscillator dissipates some of the signal power in R_{ESR}. To maintain an oscillation, the amplifier stage needs to behave as a negative resistance to compensate for the power dissipated in R_{ESR}. The value of R_{ESR} changes over temperature and may increase over time due to crystal aging. It is important that the crystal is designed properly with enough margin to ensure an oscillation at the startup as well as maintaining the oscillation over time at various temperatures. One way to measure the oscillator performance is determining the *safety factor* (SF), also known as *oscillator margin,* through a negative resistance test (Figure 4.20b). In this test, a resistor (R_Q) is placed in series with the crystal. The value of R_Q is increased incrementally to determine the largest value of R_Q with which the oscillator can start and maintain its oscillation (R_{Qmax}). The SF is defined as:

$$SF = \left(\frac{R_{Qmax}}{R_{ESR}} \right) \tag{4.16}$$

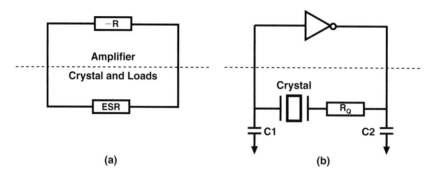

(a)   (b)

Figure 4.20: Negative Resistance Test to Determine SF

An oscillator with higher SF has a better chance of starting oscillation and maintaining its oscillation. Two types of oscillation margins are widely used. The *startup margin* is based on the largest value of R_Q that the oscillator can start oscillating. The typical value of startup SF is 5 to 15, depending on the application. The *steady state margin* is calculated from the largest value of R_Q at which the oscillator can maintain its oscillation past the startup period. An oscillator with steady state SF of 2 or less is considered risky. For noncritical-use cases, a steady state SF greater than 3 is normally sufficient.

To improve SF, you can select a crystal with lower ESR. For example, a crystal with a smaller package has less parasitic; therefore, based on Equation 4.14, the crystal has lower ESR. Similarly, reducing the parasitic of the PCB board can be helpful to enhance SF. Improving the gain of the amplifier, if possible, will improve SF as well.

4.11.2   Drive Level

The crystal can be damaged by overstressing due to excessive current. Therefore, crystal manufacturers always provide the drive level of a crystal (P_{DL} in Watts) in their datasheet. The maximum rms current of a crystal ($I_{rms\ max}$) can be calculated easily from its drive level:

$$I_{rms\ max} = \sqrt{\frac{P_{DL}}{R_{ESR}}} \tag{4.17}$$

For example, in a 16 MHz crystal with drive level of 100 μW and ESR of 50 Ω, the maximum rms current must be limited to 1.41 mA.

4.11.3   Series versus Parallel Resonant Crystals

There is no difference in the manufacturing of series versus parallel resonant crystals. The only difference is in the way the crystal is used. In a series resonant application, the crystal does not have any load capacitance. In parallel resonant use, you have to use a manufacturer-recommended load capacitance (C_{Load}) in parallel with the crystal to ensure that the resonance will occur at the specified frequency. The parallel resonant frequency is slightly higher than the series resonant frequency (e.g., less than 0.02%).

4.11.4   Crystal Frequency Tolerance

The accuracy of a crystal is measured in parts per million (ppm), which is equal to 10^{-6}. For instance, a 24 MHz crystal with ± 40 ppm accuracy can have up to ± 960 Hz of error in its oscillation frequency:

$$\pm 40 \times 10^{-6} \times 24 \times 10^{+6} = \pm 960 \text{ Hz}$$

Therefore, the center frequency of the clock generated using this crystal will be anywhere between 23.99904 MHz and 24.00096 MHz.

Normally, crystals with lower ppm are more expensive; for that reason, when selecting a transceiver, make sure it does not require a very low ppm crystal to operate properly. Many of the existing solutions can work with a crystal with ± 40 ppm accuracy.

4.11.5   Crystal Aging

The nominal frequency of the crystal may change over time even if all the external factors such as temperature remain unchanged. Manufacturers specify the aging change in ppm units. As an example, the nominal frequency of a CM309S crystal from Citizen may change ± 5 ppm over a one-year period, while the temperature is $25^\circ \pm 3^\circ$C. If a product has a long lifetime (e.g., 10 years), it is important to ensure that the crystal accuracy stays within acceptable range until the end of product lifetime.

4.11.6   Crystal Pullability

The nominal resonant frequency of the crystal oscillator is a function of the total load capacitance. Although the tuning range is very limited, it is possible to adjust the resonant frequency by modifying the load capacitance. If the shunt and motional capacitances of the crystal are known, the average frequency pulling (in ppm) for one pF change in the load capacitance can be calculated using the equation:

$$\text{Resonant Frequency change} \left(\frac{\text{ppm}}{\text{pF}} \right) = \frac{C_1 \times 10^6}{2(C_0 + C_{\text{Load}})^2} \tag{4.18}$$

For example, for $C_1 = 0.018\,\text{pF}$, $C_{\text{Load}} = 13\,\text{pF}$, $C_0 = 7\,\text{pF}$, and a resonant frequency of 8 MHz, 1 pF increase in the load capacitance reduces the resonant frequency by 22.5 ppm, or 180 Hz. It is also possible to request pullability curves from the crystal manufacturer; these illustrate the resonant frequency change for various load capacitances.

4.11.7   Crystal Overtones

A crystal may have more than one resonant frequency. The first resonant frequency is called the *fundamental frequency* of the crystal. The resonant frequencies above the fundamental frequency are known as *crystal overtones*. For instance, a designer may choose to use the third overtone of a crystal to generate the desired clock. In this case, external circuit elements (i.e., an LC tank) must be used to prevent the oscillation at the fundamental frequency. The overtone frequencies of a crystal are not exactly integer multiples of the fundamental frequency. Using the overtone frequency of a crystal instead of its fundamental frequency is a way of achieving a high-frequency accurate clock when either there is no crystal with the desired fundamental frequency or the available crystals are more expensive. The disadvantage of using a crystal in overtone mode of operation is additional external components that can increase the total board area.

If a transceiver can work with a range of crystals with various center frequencies, it will provide the flexibility of selecting the crystal with the lowest size and cost in the allowed frequency range. However, if you select any crystal with the center frequency other than the transceiver nominal crystal frequency, the total current consumption of the IC may increase because the transceiver may need to turn on additional internal circuitry to convert the oscillation frequency to the nominal clock frequency.

Also it is important that the transceiver accept an external clock in lieu of using its own crystal. In this way if a clock is already generated for another application, it can be reused for an IEEE 802.15.4 transceiver as well to save area and cost associated with an additional crystal. Also, accepting an external clock simplifies the synchronization of an IEEE 802.15.4 transceiver to any other hardware located on the same board.

4.12   Analog-to-Digital Converters

The sensor is a device that generates a signal in response to a physical stimulus such as pressure, temperature, or motion. The output of a sensor used in a wireless sensor network (WSN) is typically analog voltage or it is converted to analog voltage. For example, most commercially available accelerometers convert the acceleration in a given direction to a voltage change. This analog voltage needs to be converted to digital using an *analog-to-digital converter* (ADC) before it can be transmitted wirelessly. The transceivers developed for IEEE 802.15.4 applications normally have an integrated ADC available. The ADC can read sensor information and even move the information to memory without CPU interaction. Reading and processing the sensor information may consume a noticeable amount of power; therefore, it is recommended to gather the sensor information while the rest of the node (e.g., radio circuits) is in sleep mode. Turning everything on concurrently will increase the peak of the current drawn from the battery. Draining the battery at a high discharge rate can adversely affect battery life. In this section, basic ADC operation and performance metrics are reviewed; these can be helpful in selecting the right ADC for each application scenario.

The basic principle of analog-to-digital conversion can be described asking a simple example. Assume that we like to represent an analog voltage (V_{IN}) with only 2 bits. If the maximum possible value of this voltage is V_{REF}, the analog voltage V_{IN} can be divided by V_{REF} to normalize the maximum input voltage to 1. Figure 4.21 shows the

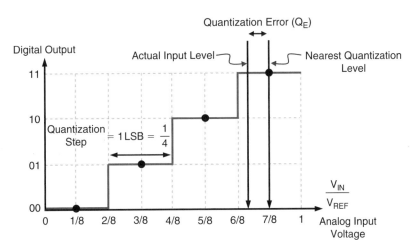

Figure 4.21: A 2-bit ADC Input/Output Relationship

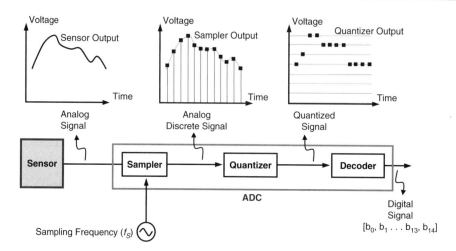

Figure 4.22: Simplified ADC Building Blocks

input/output relationship of a 2-bit ADC. The normalized analog input range is divided into four equal-sized segments and the middle value of each segment is assigned to a digital number. The digital numbers 00, 01, 10, and 11 correspond to 1/8, 3/8, 5/8, and 7/8 input levels, accordingly. These four input levels are referred to as the *quantization levels*. For example, if the analog input is any value from 1/4 to 1/2, the assigned digital number will be 01. In other words, any input number between 1/4 and 1/2 is rounded to the nearest quantization level (3/8). The *quantization error* is defined as the difference between the actual analog input value and the nearest quantization level.

Figure 4.22 is a simplified diagram of the building blocks in an ADC. The analog signal generated by a sensor is typically a low-frequency signal. The sampler captures the value of the signal at certain moments in time. The sampler output is a discrete-time signal, but the value of the signal is still analog. The quantizer rounds each sample voltage to the nearest quantized value. Finally, the decoder maps each quantized value to a digital number. If the output of the sensor is not a proper voltage, the sensor voltage should be modified to make sure that the voltage is always within 0 and V_{REF}. This is known as *signal conditioning*. It may also include filtering to remove undesired frequency components.

The maximum amount of quantization error (V_{QE}) depends on the number of bits in an ADC:

$$-\frac{V_{REF}}{2^{n+1}} < V_{QE} < \frac{V_{REF}}{2^{n+1}} \qquad (4.19)$$

where n is the number of bits. If the quantization error has a uniform probability density function, the rms value of the quantization error ($V_{QE(rms)}$) can be calculated from the following equation:

$$V_{QE(rms)} = \frac{V_{REF}}{2^n \times \sqrt{12}} \qquad (4.20)$$

The quantization error is also referred to as *quantization noise*. For example, an ideal 12-bit ADC with V_{REF} of 1.8V has a maximum quantization error of 0.22mV and $V_{QE(rms)}$ of 0.13mV. The quantization noise is present even in an ideal ADC because this noise is due to the fundamental limitation of an ADC. If the input voltage V_{IN} is a sinusoid with rms value of $V_{IN(rms)}$, the ratio of the signal energy to the quantization noise energy (SNR) at the output of an ideal ADC is the popular equation:

$$SNR = 20 \times \log_{10}\left(\frac{V_{IN(rms)}}{V_{QE(rms)}}\right) = 6.02 \times n + 1.76\,dB \qquad (4.21)$$

For instance, an ideal 12-bit ADC has a maximum SNR of 74dB. The ADC *resolution* is the number of distinct analog levels in an ADC. An *n*-bit resolution ADC has $1/2^n$ distinct analog levels.

Although Equation 4.21 is for a sinusoid input signal, it is typically used as an estimate for other types of input signals as well. In a non-ideal ADC, not only are additional sources of noise present, but also the nonlinearity of the ADC adds distortions to the ADC output signal.

The signal distortion creates undesired additional tones in the power spectral density of the quantized output signal (see Figure 4.23). In the example shown in Figure 4.23, the input analog signal is a single tone with a frequency of 5KHz. The converter is a 12-bit ADC and the ideal quantization noise level is shown at $-74\,dB_{Vrms}$. The additional noise shown in Figure 4.23 is coming from ADC circuits. Undesired spurious levels due to signal distortion are appearing in 10KHz, 15KHz, and 20KHz. The difference between the maximum spurious level and the desired signal level is known as the *spurious-free dynamic range* (SFDR). The SFDR (in dB) is an indication of ADC linearity.

The performance of a non-ideal ADC is analyzed using the ratio of the signal energy to the combined noise and distortion energy. The *signal-to-noise-and-distortion*

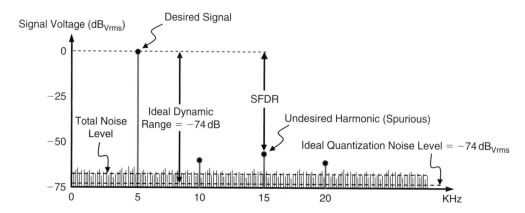

Figure 4.23: Example of an Output Signal Spectrum in a 12-bit (Nominal) ADC

(SINAD) ratio, sometimes abbreviated as signal-to-noise-and-distortion ratio (SNDR), is determined using the following equation:

$$SINAD = 20 \times \log_{10}\left(\frac{\text{Signal Voltage (rms)}}{\text{Total Noise Voltage (rms)} + \text{Total Distortion Voltage (rms)}}\right) dB$$

(4.22)

In practice, the SINAD is used instead of the SNR as a measure of signal quality.

The noise and distortion prevent an ADC from delivering its nominal (ideal) resolution. As the noise and distortion levels are increased, some of the least significant bits of the ADC may reflect only the noise instead of any valuable signal information. The effective number of bits (ENOB) is the actual ADC resolution, when the noises and distortions are taken into account. The ENOB of an ADC is determined using the SINAD:

$$ENOB = \frac{SINAD - 1.76}{6.02}$$

(4.23)

For example, in an ADC with 12-bit nominal resolution, if the SINAD is equal to 59 dB, the ENOB will be 9.5 bits. The ENOB should be used to compare the performance of different ADCs instead of the ADC nominal (theoretical) number of bits.

Another performance metric in an ADC is the *conversion time*. The conversion time is the time it takes for the ADC to complete the conversion of an analog signal to digital. Generally, when an ADC is designed, there is a trade-off between the number of bits (resolution) and the conversion time. The resolution can be improved at the expense of longer conversion time. Alternatively, the conversion time can be improved by reducing the ADC resolution. The analog output of a sensor is typically low frequency (i.e., varies slowly). Therefore, high-resolution (e.g., 12 to 14 bits) ADCs with long conversion time (e.g., several microseconds) are common in wireless sensor networks.

The ADC bandwidth is the maximum frequency at which the ADC can operate efficiently and digitize an analog signal. In many wireless sensor networking applications, bandwidth of several hundred KHz is sufficient. The required sampling frequency in an ADC (Figure 4.22) depends on the analog signal bandwidth. If the signal bandwidth is B, the Nyquist sampling theorem indicates that the sampling frequency (f_S) must be at least twice the signal bandwidth to ensure that the signal can be reconstructed properly after the sampling. The choice of the sampling frequency divides the ADCs into two general types: Nyquist-rate ADCs and oversampling ADCs. The Nyquist-rate ADCs sample the signal only fast enough that the signal can be reconstructed properly later. Although based on the Nyquist theorem, f_S can be as low as $2 \times B$; the sampling frequency f_S is normally selected 2 to 6 times higher than the minimum required by the Nyquist theorem for practical considerations. The oversampling ADCs, on the other hand, select a sampling frequency that is much (e.g., 100 times) higher than $2 \times B$. This very high oversampling allows these ADCs to improve the SNR of the output signal and effectively improve the ADC resolution. One popular oversampling ADC is the *sigma-delta* ADC, which uses the combination of oversampling and a noise-shaping (filtering) mechanism to reduce the noise energy in the frequency band of interest and improve the SNR accordingly.

The clock used for the sampling stage of an ADC is not ideal and contains *jitter*, the undesired random variation of the width of the pulse in a clock from one period to another. In other words, there is a small but random variation in the rising and falling edges of the clock. The clock jitter can degrade the output signal SNR and consequently reduce the ENOB in an ADC. The ADC's architecture determines the sensitivity of the ADC's performance to clock jitter. Generally, the higher the input analog signal frequency, the tighter the requirement for clock jitter.

References

[1] D. M. Dobkin, "*RF Engineering for Wireless Networks*," Elsevier Science, 2005.
[2] "IEEE 802.15.4: Wireless Medium Access Control (MAC) and Physical Layer (PHY) Specifications for Low-Rate Wireless Personal Area Networks (WPANs)," Sept. 2006.

[3] D. A. Jones and K. Martin, "*Analog Integrated Circuit Design*," John Wiley & Sons, 1997.

[4] W. C. Y. Lee, "*Mobile Communication Engineering, Theory and Applications*," 2nd ed., McGraw-Hill, New York, 1998.

[5] J. G. Proakis, "*Digital Communications,* " 4th ed., McGraw-Hill, 2000.

[6] T. S. Rappaport, "*Wireless Communications: Principles and Practice*," Prentice Hall, NJ, 2002.

[7] B. Razavi, "*RF Microelectronics*," Prentice Hall/PTR, Nov. 1997.

CHAPTER 5

RF Propagation, Antennas, and Regulatory Requirements

Understanding the propagation characteristics of an environment is essential in properly implementing and troubleshooting wireless networks. This chapter covers the fundamentals of RF propagation, range estimation, and range improvement techniques. Basic characteristics of antennas are also reviewed because selection of the antenna can have a considerable impact on the overall performance, range, and size of a wireless node. Local governments issue regulations to help maintain wireless communication system integrity and product performance in their regions. Regulatory requirements in North America, Europe, and Japan are summarized at the end of this chapter.

5.1   Path Loss

In establishing any wireless network, one of the basic questions that must be answered is how far two wireless nodes can be placed from each other while maintaining a reliable wireless connection. The answer depends on many parameters, including receiver sensitivity, transmitter output power, signal frequency, and signal propagation environment.

Consider an isotropic antenna (an antenna that emits a signal uniformly in all directions) located in a place where no other materials or emitted signals are present except the signal emitted evenly from the antenna in all directions. This is called signal propagation in *free space*. The signal propagation can be imagined as an expanding sphere centered at the antenna (Figure 5.1). In free space, the signal power at distance d from the antenna is proportional to P_0/d^2, where P_0 is the signal power at the antenna. If the distance from the antenna is doubled, the signal power will be reduced by a factor of four.

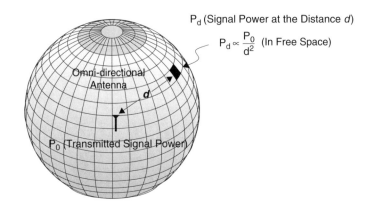

Figure 5.1: Omnidirectional Antenna Propagation in Free Space

The signal power at distance d is also a function of frequency. In free space, the signal power at distance d can be calculated from the *path-loss* equation:

$$P_d = P_0 - 10 \times 2 \times \log_{10}(f) - 10 \times 2 \times \log_{10}(d) + 27.56 \tag{5.1}$$

where:

P_d is the signal power (in dBm) at distance d
P_0 is the signal power (in dBm) at zero distance from the antenna
f is the signal frequency in MHz
d is the distance (in meters) from the antenna

As can be seen, increasing the signal frequency reduces the signal power at distance d. For example, if the antenna transmits a 0 dBm signal at 914 MHz, the signal power at 10 meters' distance from the antenna will be around -52 dBm, whereas if you keep the transmit signal power the same and just increase the signal frequency to 2450 MHz, the signal power at 10 meters from the antenna will be reduced to -60 dBm.

In most real-world scenarios, such as inside a typical house, the preceding free-space equation will not be accurate enough. This is because parts of the transmitted signal will be absorbed in different surrounding materials, the signal may be reflected multiple times from various objects, and delayed versions of the signal may be added to the original signal. All these incidents will change the signal power. Based on various experiments [1], Equation 5.1 can be modified to the following:

$$P_d = P_0 - 10 \times n \times \log_{10}(f) - 10 \times n \times \log_{10}(d) + 30 \times n - 32.44 \tag{5.2}$$

Table 5.1: Path-loss Exponent (*n*) for Different Environments

N	Environment
2.0	Free space
1.6 to 1.8	Inside a building, line of sight [2]
1.8	Grocery store [2]
1.8	Paper/cereal factory building [2]
2.09	A typical 15 m × 7.6 m conference room with table and chairs [3]
2.2	Retail store [2]
2 to 3	Inside a factory, no line of sight [2]
2.8	Indoor residential [4]
2.7 to 4.3	Inside a typical office building, no line of sight [1]

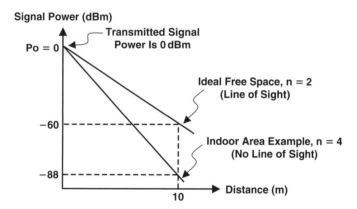

Figure 5.2: Effect of the Environment on Signal Power at a 10-meter Distance from the Antenna

The value of *n*, which is called the *path-loss exponent*, is determined experimentally (Table 5.1).

Line of sight means that there is an unobstructed path from the transmitter to the receiver. Figure 5.2 shows the effect of the value of *n* on the signal power at 10 meters from the antenna. For a frequency of 2450 MHz and transmitted signal power of 0 dBm, *n* = 2 results in a −60 dBm signal power whereas *n* = 4 yields only a −88 dBm signal.

5.2 Signal Wavelength

The signal propagation speed in air is approximately the speed of light ($C = 3 \times 10^{+8}$ meters per second). The signal wavelength is the distance that wave travels (in meters) during one signal period (T):

$$\text{Signal wavelength } (\lambda) = C \times T = \frac{C}{f}$$

For example, a 2485 MHz signal has a wavelength of around 12 cm (4.7 inches).

5.3 Signal Penetration

When a signal is penetrating an object, the absorption characteristic of the object material and its temperature as well as the signal frequency will determine how much the signal will be attenuated. Each material is associated with an attenuation constant, a (dB/m), which is a function of the temperature and the signal frequency. For example, the water attenuation constant at room temperature for a 2.4 GHz signal is around 330 dB/m, which means that one foot of water will attenuate the signal around 100 dB. The human body consists of about 70% water and therefore attenuates the RF signals significantly. Metals, in general, do not allow any signal penetration and reflect almost the entire signal.

Sometimes the attenuation constant is given as α (Nepers/m) instead of a (dB/m). Similar to dB, Neper is a unit in logarithmic scale. Neper uses base $e(\approx 2.718)$ logarithm, while decibel (dB) is using base-10 logarithm. To convert Nepers/m to dB/m you can simply use the equation:

$$a\left(\frac{dB}{m}\right) = 8.68 \times \alpha\left(\frac{\text{Nepers}}{m}\right) \tag{5.3}$$

For example, when α is equal to 4.5 (Nepers/m), a is 39 dB/m.

The effect of signal frequency on penetration depends on the type of material. Generally speaking, in most frequency ranges of interest and considering the types of materials that can be found in a house or an office building, when the signal frequency is increased, the signal attenuation will increase. In other words, lower-frequency signals penetrate most materials better compared to higher-frequency signals.

Signal penetration of an object depends on the angle at which the signal hits the surface of the object. Consider the gypsum wall (drywall) in Figure 5.3, which is commonly used in construction as an interior wall and consists of two gypsum boards and light gauge

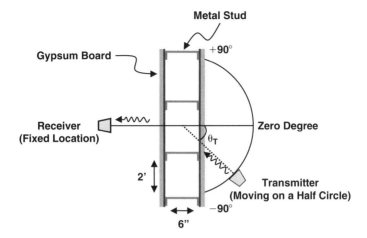

Figure 5.3: Measuring Signal Penetration of a Gypsum Wall

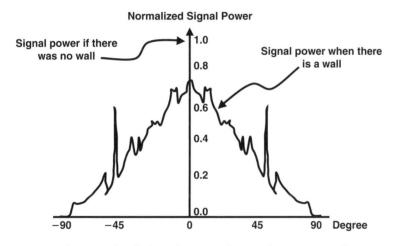

Figure 5.4: The Received Signal Power for Various Transmitter Angles Shown in Figure 5.3

steel joints in between. The signal hits the wall at an angle of θ_T. The relative angle of the transmitted signal and the wall changes by moving the transmitter on a half-circle path. The received signal power is measured at the other side of the wall. In this way, we can generate a graph similar to Figure 5.4 that shows the received signal power for different transmitter positions [5]. Figure 5.4 shows that the maximum received signal happens when the signal hits the wall at a 0° angle and the amount of the received signal approaches zero as the transmitted signal angle approaches ±90°.

Table 5.2: Signal Attenuation in Various Objects

Object (at Room Temperature)	Signal Frequency	Signal Attenuation (dB)
Soft cloth partition wall (2 inches) [1]	914 MHz	1.5
Building floor [1]	914 MHz	17
Building floor [6]	1–2 GHz	23
Interior concrete wall (4 inches) [6]	1–2 GHz	6
Interior brick wall (5 inches) [6]	1–2 GHz	2.5
Plaster board [6]	1–2 GHz	1.5
Reinforced glass [6]	1–2 GHz	8

Several experimental measurements have captured the approximate signal attenuation caused by various objects. For instance, some experiments [1] showed that a soft cloth partitioning wall, which is common in office areas, will attenuate a 914 MHz signal around 1.5 dB. In a multistory office building, if the receiver and the transmitter are on two different floors, a 914 MHz signal is shown to be attenuated around 17 dB as it penetrates each floor [1].

Table 5.2 summarizes the results of some of these experiments. These are ballpark estimates and may vary considerably from one experiment to another.

5.4 Reflection, Scattering, and Diffraction

Consider the example shown in Figure 5.5. A moisture sensor communicates with an irrigation controller unit reporting the soil moisture level at a certain depth. The controller will use this information along with the ambient temperature, time, and date to determine the watering period. In this example, there are at least four different ways for the transmitted signal to reach the receiver. The first path in this figure represents the signal penetration through the house. (Signal penetration and absorption were discussed earlier, in Section 5.3.) The second way is through *diffraction*. Diffraction occurs when the path of an electromagnetic wave is blocked by an obstacle with relatively sharp edges. The sharp edge in this example is the corner of the house. This sharp edge will cause the signal to bend toward the object and as a result, the signal may reach the receiver, although there is no line of sight between the receiver and the transmitter.

Whenever an electromagnetic wave is incident on a surface, a portion of or the entire signal will be reflected. Metal reflects almost the entire signal. A thin slab of plywood, in

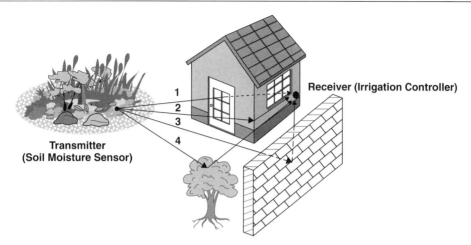

Figure 5.5: The Transmitted Signal can Reach the Receiver by Way of (1) Penetration, (2) Diffraction, (3) Reflection, or (4) Scattering

comparison, reflects about 60% of the signal power. Path 3 in Figure 5.5 shows that the signal is reflected from a block wall and reaches the receiver.

Path 4 represents the *scattering* incident. Scattering occurs when an electromagnetic wave is incident to a rough surface. The roughness of the surface depends on the signal wavelength. Generally speaking, if variation on a surface is more than $\lambda/8$, the surface is considered to be rough and will scatter the signal in many directions [7]. For example, if the signal is at 2450 MHz, the wavelength is about 4.8 inches (12.2 cm), and variations of more than 0.6 inches (1.5 cm) on a surface, which is the case in a typical tree, will cause the signal to be scattered in many directions. There are a number of studies on tree-scattering [8] that provide an approximate propagation model based on experimental measurements.

5.5 Multipath Environment

In the example of Figure 5.5, we observed that the transmitted signal may find several different paths to the receiver due to reflections, diffractions, and scatterings. Let's start with only two paths, as shown in Figure 5.6; the only medium present is air. The propagation delay is the time it takes for the signal to reach to the receiver. Given that the distances of d_1 and d_2 (in meters) are not equal, the received signals from paths 1 and 2 will have different delays; therefore, their phases will be different as well. The

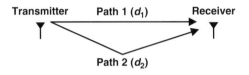

Figure 5.6: The Simplest Multipath Scenario

Table 5.3: Average Delay Spreads

Environment	Delay Spread (ns)
Indoor residential (no line of sight) [4]	70
Indoor office (no line of sight) [4]	100
Indoor commercial (no line of sight) [4]	150
Large open space (indoor and outdoor; no line of sight) [9]	150 to 250

propagation delay is equal to d/C. For example, if d_1 and d_2 are 10 m and 15 m, received signal delay from path 1 and 2 will be 33 ns and 50 ns, respectively.

Typical average delay spread values for some common indoor environments are provided in Table 5.3.

The phase difference (in radians) between the received signals can be calculated from:

$$\Delta\theta = \frac{2\pi f}{C}(d_1 - d_2) = \frac{2\pi f}{C}\Delta d \qquad (5.4)$$

Therefore, unlike the delay, the phase difference is a function of the signal frequency. If the signal frequency is 2450 MHz, one foot (30.48 cm) of difference between the paths will result in $4\pi + 3$ radians of phase difference.

In most practical scenarios, a node receives the summation of several versions of the transmitted signals due to multiple reflections, diffractions, and scattering. These signals have different delays and phase shifts; therefore, the summation will be a distorted signal. This is called the *multipath distortion*. Adding different versions of the signal may increase the signal strength, but the signal quality may become poor. A low-quality signal can result in poor communication.

5.5.1   *Multipath-Induced Additional Random Phase*

The bad news is that the additional phase shift due to multipath can be any random phase between $-180°$ and $180°$. The good news is that if the receiver and the transmitter are not moving and the environment is stationary, this additional random phase is a constant value.

In the differential phase detection method, the information is in the phase difference of two consecutive symbols rather than the absolute phase values. Therefore, an undesired constant additional phase will not impact a receiver that uses the differential phase detection method. Another approach to coping with the multipath constant additional phase is to estimate this constant value and then adjust the receiver accordingly to compensate for it.

5.5.2   *Multipath Null*

In a multipath environment, several versions of the same signal with different phases, delays, and attenuations will be added together at the receiver location, so there is always the possibility that at some locations, the signals could cancel each other out almost entirely. Therefore, regardless of the transmitted signal strength, it is impossible or very difficult to receive any data. Consider the example in Figure 5.6, where there are only two paths. From Equation 5.4, it is easy to show that if the difference between the two paths is an odd multiple of the signal wavelength, the received signals will have 180° phase difference and can reduce the signal power significantly.

$$\Delta d = (2K + 1)\frac{\lambda}{2} \Rightarrow \text{ The signals will have } 180° \text{ phase difference}$$

One way to overcome the multipath issue is to use the *receiver antenna diversity* technique. In this method, two antennas are used instead of one in the receiver. The recommended distance between the antennas is equal to the signal wavelength (λ). This way if one antenna is in a multipath null (also known as *deep-fading region*), the other antenna has a good chance of being outside the deep-fading region. The receiver can switch between these two antennas to escape from a multipath null. Placing the diversity antennas less than $\lambda/2$ of each other significantly reduces the chance of overcoming a multipath null. Also, the antenna separation distance should not exceed 4λ.

The antenna diversity can be implemented on the transmitter side as well. This way the transmitter node can switch between two antennas to overcome a possible multipath issue. For the best result, both the transmitter node and the receiver node can be equipped with diversity antennas.

5.5.3   Fading Channel and Fade Margin

In a stationary environment that does not have any multipath, the only nonideality is some additional wideband (white) noise. This model is called the *additive white Gaussian noise* (AWGN) channel. But this model will not be accurate when multipath exists; the *fading channel* model should be used instead. The fading channel, unlike the AWGN channel, acts as a frequency-selective filter and can attenuate some of the frequency components of the transmitted signal. If this attenuation is very high, the signal is said to be in *deep fade* (also known as *multipath null*).

In designing a wireless system, it is sometimes necessary to leave some margin to account for the fading effect. This is called *fade margin* and its value is determined experimentally. For instance, if the receiver sensitivity is -95 dBm and the recommended fade margin for a particular environment is 8 dB, the effective sensitivity level of -87 dBm should be used in all system-level calculations. A fade margin of 6 dB to 10 dB is typical for indoor applications.

5.5.4   Effect of Frequency Channel on Multipath Performance

Changing the frequency will change the fading characteristic and generally can help bring a receiver node out of multipath null. To know the reason, we need to understand the *coherence bandwidth* concept. Coherent bandwidth is a measure of how much the frequency can be changed while experiencing a similar fading environment. The amount of coherent bandwidth can be approximated by:

$$\text{Coherence bandwidth } (\text{B}_\text{C}) \text{ in Hertz} = \frac{1}{2\pi \times (\text{Delay Spread})} \tag{5.5}$$

For instance, in an indoor residential environment (Table 5.3), the delay spread is 70 ns, which means that the coherence bandwidth is 2.3 MHz. This value of the coherence bandwidth implies that if the receiver is in deep fade and the carrier frequency is shifted to 1 MHz away from its current value, the receiver is likely to stay in deep fade. But if the carrier frequency is shifted to 10 MHz away, the receiver will potentially experience a different fading environment and can have a better chance of receiving the signal.

5.5.5   Effect of Signal Spreading on Multipath Performance

If the signal frequency bandwidth is smaller than the coherent bandwidth, the entire signal spectrum will experience similar fading. This means that if a portion of this signal

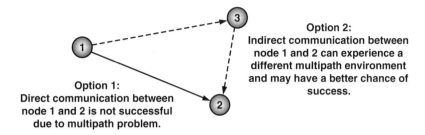

Figure 5.7: Overcoming the Multipath Issue Using an Alternative Route

spectrum is in deep fade, it is likely that the entire signal spectrum will be in deep fade. For example, in a large outdoor area with delay spread of 250 ns, the coherence bandwidth (B_C) is 637 kHz. If there is no spreading and the signal bandwidth (B_S) is 250 KHz, then $B_S < B_C$ and the entire 250 KHz spectrum of the signal experiences the same fading. If the receiver happens to be in a multipath null, the entire signal spectrum is in deep fading.

Signal spreading increases the signal bandwidth, and when the signal bandwidth becomes sufficiently larger than the coherence bandwidth, it is possible that when a portion of the signal spectrum is in deep fade the rest of the signal will experience a different, and potentially better, fading environment. In our earlier outdoor example, if the signal bandwidth (B_S) is increased to 2 MHz using signal spreading, the signal bandwidth exceeds the coherent bandwidth and a portion of the signal spectrum is always outside any possible multipath null. A signal with partly disturbed spectrum may still have a chance of recovery by the receiver.

5.5.6 Mesh Networking to Improve Multipath Performance

The fading characteristic of the communication channel depends on the location of both receiver and transmitter; choosing an alternative path can potentially help overcome an existing multipath issue. This idea is shown in Figure 5.7, where the signal from node 1 cannot reach node 2 directly, but when the alternative path is used, the signal is transmitted from a different location (node 3) and experiences a different fading environment, which could have a better chance of reaching node 2.

5.6 Doppler Frequency Shift

It is unlikely for the Doppler frequency shift to have an impact in a typical ZigBee application, but its concept is explained here for your reference.

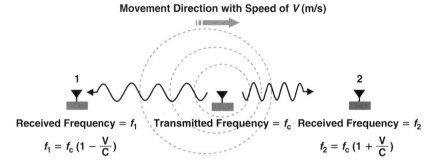

Figure 5.8: Doppler Frequency Shift

Some ZigBee applications require communication with a transceiver that is located on a moving object. Figure 5.8 shows a transmitter that is moving away from receiver 1 and toward receiver 2 with the speed of V(m/s). If the transmitter was not moving (V = 0), the received signal frequency at receivers 1 and 2 would be the same as the original transmitted frequency (f_C). But when the transmitter is moving, due to a phenomenon first described by mathematician and physicist Johann Christian Doppler, the frequency received by receiver 2 will become higher than f_C:

$$f_2 = f_C \left(1 + \frac{V}{C} \right)$$

where C is the speed of light in meters per second. Similarly, the signal frequency received by receiver 1 will be lower than the original transmitted frequency (f_C, in Hz):

$$f_2 = f_C \left(1 - \frac{V}{C} \right)$$

This means that in communicating with a moving object, there is a frequency error (Δf) that is proportional to the speed of the moving object:

$$\Delta f(ppm) = \frac{V}{C} \times 10^6 \tag{5.6}$$

$$\Delta f(Hz) = \frac{V}{C} \times f_C \tag{5.7}$$

For instance, in a personnel-tracking system, a person wearing a transmitter is walking toward a fixed receiver with a typical walking speed of 4 miles/hour (about 0.7 m/s). In that case the frequency shift due to Doppler will be 0.002 ppm. Considering that the receiver is supposed to tolerate ± 40ppm of frequency error, the frequency shift due to Doppler is insignificant in this example.

Alternatively, we can analyze the effect of moving objects using the coherence time (T_C). Coherence time is the time duration over which the channel time-domain variations are negligible. The approximate value of the coherence time can be calculated from the Doppler frequency shift (Δf):

$$T_C = \frac{1}{\Delta f} \quad \text{(Seconds)} \tag{5.8}$$

where Δf is the Doppler frequency shift in Hz.

In our personnel-tracking example, assuming that the frequency of operation is 2450 MHz, the coherence time will be equal to 175 ms. This coherence time is much longer than any typical IEEE 802.15.4 packet length. In this example, during the transmission of any packet, the channel characteristics stay almost unchanged. Therefore, the effect of Doppler frequency shift can easily be neglected.

Since the electromagnetic wave propagation speed is much higher than the speed of most moving objects in a typical ZigBee wireless network, the Doppler frequency shift is practically negligible in most ZigBee applications.

5.7   Site Survey

A *wireless site survey* is a physical survey of the premises where the wireless network will be installed. A site survey report helps visualize the wireless network coverage areas and data rates. Commercial tools are available that use actual measurements to extrapolate and create a coverage graph that shows signal strength and signal quality at different locations of the site. In the simplest form of site survey, you can walk through the facility with a handheld spectrum analyzer and measure the signal strength and possible interferences. There are also prediction tools that use the building floor plans and wireless node proposed locations to predict the coverage mathematically instead of using actual measurements.

5.8 Range Estimation

From the concepts covered in previous sections, it is easy to come up with an equation to estimate the range associated with two wireless nodes. In Figure 5.9, if node A is a transmitter and node B is a receiver and the following values are known:

- Node A transmit power (including the antenna gain, if any) = P_o(dBm)

- Node B receiver sensitivity = P_r(dBm)

- Path-loss exponent (use Table 5.1) = n

- Fade margin (see Section 5.5.3) = F_m(dB)

- Signal frequency = f(MHz)

Then the estimated range (R, in meters) can be calculated from the following equation:

$$R = 10^{\left(\frac{P_o - F_m - P_r - 10 \times n \times \log_{10}(f) + 30 \times n - 32.44}{10 \times n} \right)}$$

(5.9)

For example, if node A transmits a 2450 MHz signal with 0 dBm output power and receiver sensitivity is -92 dBm in an environment with a path-loss exponent of 2.8 and a fade margin of 10 dB, the estimated range is 24 meters (79 feet). This book's companion Website has a simple calculator for range estimation.

5.8.1 Range Improvement Techniques

There are at least three ways to increase the range of wireless nodes. Equation 5.9 suggests that increasing the output power and/or improving sensitivity will extend the range. Alternatively, the range can be extended using a node-hopping (mesh networking) technique.

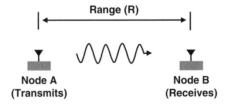

Figure 5.9: Range Estimation

5.8.1.1 Range Improvement Using External PA and/or LNA

The output power of a typical IEEE 802.15.4 transmitter is normally adjustable and the range can be increased by increasing the transmitter output power. But if the maximum output power of the transmitter, which is typically around $+3$dBm, is not sufficient for the application, an external power amplifier (PA) can be added to boost the output power and consequently improve the range. Figure 5.10a shows a simplified diagram of how an external PA can be added in the transmit path. Transceiver ICs normally have a pin (called a T/R pin) that indicates whether the transceiver is in transmit or receive mode. This T/R pin can be used to put the external PA in the transmit path during the transmit mode of operation and remove it from the path when the transceiver is in receive mode.

The drawbacks of this method are additional cost of an external PA and higher current consumption of the node due to an extra PA. An extra PA can significantly reduce the battery life of a wireless node. Also, an additional PA not only amplifies the desired output signal, it also magnifies the unwanted out-of-band emissions. Therefore, it is recommended that you reexamine the conformity of the wireless node to regulatory requirements such as FCC and ETSI after an extra PA is added to a node. In most cases, adding an additional filter after the external PA to suppress the out-of-band emissions might be necessary. Two additional T/R switches in the receive path can degrade the receiver sensitivity around 1dB

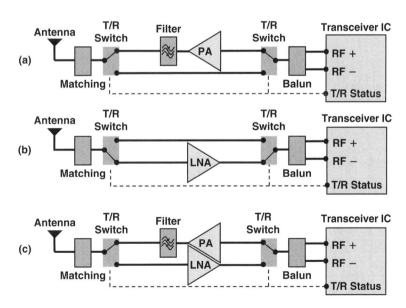

Figure 5.10: Range Extension Using External LNA and/or PA

to 2dB, depending on the type of T/R switches used. In some of the existing transceiver ICs, the balun in Figure 5.10a is inside the transceiver IC package.

As an example, using Equation 5.9, by adding an external PA to node A in Figure 5.9 we increase the output power to $+10\,$dBm from $0\,$dBm and consider a 2450 MHz signal, whereas node B receiver sensitivity is $-92\,$dBm. In an environment with a path-loss exponent of 2.8 and a fade margin of 10 dB, the estimated range will increase to 54.7 m (179 feet) from 24 m (79 feet).

$$R = 10^{\left(\frac{P_o - F_m - P_r - 10 \times n \times \log_{10}(f) + 30 \times n - 32.44}{10 \times n}\right)}$$

$$= 10^{\left(\frac{+10 - 10 - (-92) - 10 \times 2.8 \times \log_{10}(2450) + 30 \times 2.8 - 32.44}{10 \times 2.8}\right)}$$

$$R = 54.7\,m$$

The concept of Signal to Noise Ratio (SNR) was discussed in chapter 4. The SNR is an indication of the signal quality and increasing SNR will improve receiver Packet Error Rate (PER). The RF and analog blocks in a receiver degrade the SNR of the received signal as they pass the signal to the receiver digital block. Noise Figure (NF) is a measure of SNR degradation caused by a block. Noise Figure of a block is the ratio of the SNR at the input of the block to the SNR at the output of the block. In an ideal case, with no SNR degradation, this ratio is equal to 1. The Noise Figure is generally provided in dB scale. If the NF is provided as a straight ratio instead of decibel, it is referred to as Noise Factor.

Another method for increasing the range is improving the sensitivity using an additional low-noise amplifier (LNA) in the receiver path (Figure 5.10.b). An LNA can improve overall NF of a receiver and consequently, improve the receiver sensitivity and range. Both the gain and the noise figure (NF) of this additional LNA impact the overall receiver sensitivity. To calculate the sensitivity improvement, you need to have the following:

- Original receiver sensitivity (without the additional LNA) = P_r(dBm)

- Original receiver NF (without the additional LNA) = NF_r(dB)

- External LNA NF = NF_{LNA}(dB)

- External LNA gain = G_{LNA}(dB)

- Total loss due to each T/R switch = $L_{T/R}$(dB)

Normally, the receiver NF is not quoted in the transceiver IC datasheet, but the manufacturer may provide the receiver NF value on request, or you can estimate the

NF using the receiver sensitivity value. The improved sensitivity level of the receiver (P_{r+LNA}) can be calculated from the equation:

$$P_{r+LNA} = 10 \times \log_{10} \left(10^{NF_{LNA}/10} + \frac{10^{NF_r/10} - 1}{10^{G_{LNA}/10}} \right) - NF_r + L_{T/R} + P_{r+LNA} \qquad (5.10)$$

After calculating the new receiver sensitivity level, Equation 5.9 will provide an estimate of the improved range. For example, if in Figure 5.10b the original receiver sensitivity is −92 dBm with a noise figure of 17 dB, the addition of two T/R switches with total loss of 1 dB and an external LNA with gain of 15 dB and a noise figure of 1.5 dB will improve the overall sensitivity to −103 dB:

$$P_{r+LNA} = 10 \times \log_{10} \left(10^{1.5/10} + \frac{10^{17/10} - 1}{10^{15/10}} \right) - 17 + 1 - 92 = -103.3 \, \text{dBm}$$

If a transmitter is emitting a 2450 MHz signal with 0 dBm output power in an environment with a path-loss exponent of 2.8 and fade margin of 10 dB, the estimated range will improve to 60.7 m (199 feet) from 24 m (79 feet).

Intuitively it is possible to use both external LNA and PA to maximize the range (Figure 5.10c). This book's companion Website has a simple calculator for range extension methods discussed in this section.

5.8.1.2 *Mesh Networking to Improve Range*

This is one of the important advantages of establishing a wireless network using ZigBee. In a mesh network, nodes are interconnected with other nodes so that at least two pathways connect each node to the rest of the network. As illustrated in Figure 5.11, node A uses nodes B, C, and D as *routers* to send the message to a faraway node E, which is outside the direct reach of node A. The details of mesh networking were covered in Chapter 4.

5.9 Antenna Selection Considerations

Any wireless device has an antenna that converts the electric currents generated by the transceiver circuits into electromagnetic waves, and vice versa (see Figure 5.12). Antennas come in different shapes, sizes, gains, and impedances. Selecting the right antenna for an application can have a considerable impact on the overall performance,

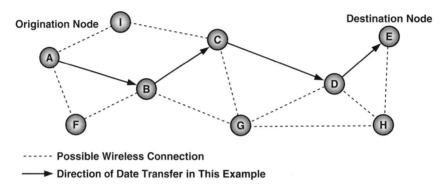

Figure 5.11: Mesh Networking Can be Used to Extend the Range

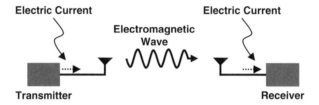

Figure 5.12: An Antenna Converts Electric Currents into Electromagnetic Waves, and Vice Versa

range, and size of the wireless nodes. This section reviews the basic principles of antennas as well as characteristics of a number of antennas.

5.9.1 Antenna Gain

An *isotropic antenna* (also known as an *omnidirectional antenna*) emits the signal uniformly in all directions. In other words, at distance *d* from the antenna, in any direction, the transmitted signal power is the same (Figure 5.1). Although building a truly omnidirectional antenna is not feasible, this ideal antenna is frequently used to simplify range estimation analysis as well as provide a reference point in comparing different types of antennas.

In a *directional antenna*, in contrast, the signal strength at distance *d* from the antenna in some directions is stronger than in other directions. Antenna *gain* is the ratio of the signal strength in the direction of strongest radiation to that of an ideal isotropic antenna. An antenna gain unit is dB*i* (*i* stands for *isotropic*). For example, a directional antenna with gain of 2.5 dB*i* will increase the output signal power in one direction by 2.5 dB compared to the output signal power of an isotropic antenna.

Although a directional antenna can increase the signal power in one direction, the total radiated signal power of any antenna cannot exceed that of an ideal isotropic antenna. Therefore, a directional antenna with positive gain in certain directions will naturally have negative gain in other directions to reduce the total emitted power below that of an ideal isotropic antenna.

Antenna gain is the same during receive and transmit modes. For example, if antenna gain is $+3$ dB*i* in the direction that the signal is received, the received signal power is 3 dB stronger compared to an omnidirectional antenna.

Antenna gain varies with frequency; each antenna is designed, or tuned, to deliver its maximum gain in one or more frequency bands. The antenna gain normally drops at higher frequencies and can help meeting regulatory emission requirements, as discussed in Section 5.10. This is because most transmitters not only broadcast the desired signal, they also transmit undesired higher-frequency spurs. Low antenna gain at high frequencies helps attenuate these undesired signals.

5.9.2 Antenna Radiation Pattern Graphs

One of the very useful graphs for analyzing and comparing different antennas is the *radiation pattern graph*. A simplified example of a radiation pattern is shown in Figure 5.13. The antenna itself is placed at the center of the graph and the antenna gain is measured at various angles. For an ideal isotropic antenna, the antenna gain is the same at any angle; therefore, the gain is a circle with 0 dB*i* value. For any practical antenna, the antenna gain varies at different angles from the antenna. The gain can exceed 0 dB*i* at some angles, but it would fall below 0 dB*i* at other angles to ensure that the summation of the antenna gain at all angles does not exceed the gain of an ideal isotropic antenna. Also remember that radiation pattern changes with frequency and the graph must be plotted for the frequency of operation.

The antenna radiates in all three dimensions, whereas the graph in Figure 5.13 is only two dimensional. To compensate for this, normally three antenna radiation pattern graphs from different directions are provided for one antenna (see examples on the book's companion Website).

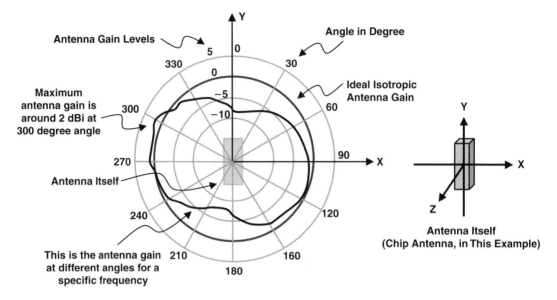

Figure 5.13: Example of an Antenna and its Radiation Pattern

5.9.3   Antenna Radiation Efficiency

An antenna receives electric signals from the transmitter circuits and converts them into electromagnetic waves. The efficiency of the antenna in performing this conversion, known as *antenna radiation efficiency*, is defined as the ratio of the power dissipated into space to the net power delivered to the antenna by the transmitter circuits.

During the receive mode of operation, an antenna converts the electromagnetic waves into electric signals. The efficiency of this operation is the same as the antenna radiation efficiency.

Antenna radiation efficiency is taken into account when antenna gain is calculated. Therefore, many antenna manufacturers choose to provide only antenna gain information, which is sufficient for most applications.

5.9.4   Antenna Impedance

Although it is a fascinating phenomenon that an antenna converts electric currents to electromagnetic (EM) waves, in many analyses we can simply replace the antenna with an equivalent impedance (for example, a 50 Ohm resistor) connected to ground (Figure 5.14) and not directly deal with EM properties of the antenna. For off-the-shelf antennas, manufacturers specify the antenna impedance in its datasheet. In a custom antenna, however,

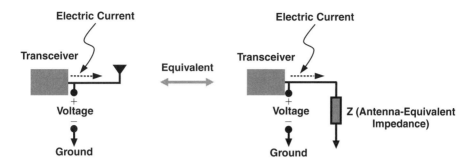

Figure 5.14: An Antenna can be Replaced by an Equivalent Impedance in Most Analyses

the impedance needs to be measured or simulated. Antenna impedance is the same during receive and transmit modes. Antenna impedance is also known as *radiation resistance*.

Transceiver manufacturers always specify the antenna impedance that will be optimal for their transceiver ICs. Many IEEE 802.15.4 transceiver manufacturers suggest using antennas with either 50 or 200 Ohm equivalent impedances. The antenna-equivalent impedance changes with frequency, and because a transceiver is designed to perform optimally with a certain antenna impedance, it is important to know whether or not the antenna impedance is close enough to the manufacturer-recommended value. To quantify close enough, the *voltage standing wave ratio* (VSWR) figure of merit is defined.

VSWR is a measure of how efficiently radio-frequency power is transferred from a power source into a load. When there is an impedance mismatch between the source and the load, a portion of the signal bounces back and creates an undesired standing wave. VSWR is the ratio of the maximum to the minimum of this undesired signal wave. In an ideal scenario, VSWR is equal to 1, which means that there is no undesired standing signal. Without getting into detail, it is possible to use VSWR to evaluate antenna impedance using Table 5.4.

The generic graph of VSWR versus frequency in Figure 5.15 shows that the antenna VSWR is less than 2.0 in the entire 2.4–2.5 GHz band. Therefore, if this antenna is used with a 2.4 GHz ISM band transceiver, the output power and the receiver sensitivity degradation due to antenna mismatch are less than 0.5 dB.

5.9.5 Power Transfer Efficiency

The efficiency of an antenna in converting EM waves into electric signals was discussed in Section 5.9.3. The next logical step is determining the power transfer efficiency from the antenna into the receiver.

Table 5.4: VSWR

VSWR Value	Meaning
1	Antenna impedance is exactly equal to the nominal value (e.g., 50 Ohms).
1 to 2	Antenna impedance is close enough to the nominal value. VSWR of less than 2.0 corresponds to less than 0.5 dB degradation in output power or receiver sensitivity.
Above 2	It is not recommended to use an antenna with VSWR of above 2 due to degradation in output power and sensitivity performance of the transceiver.

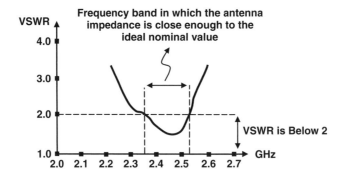

Figure 5.15: Example of an Antenna VSWR Versus Frequency

The amount of signal power that the receiver will actually receive from the antenna will depend not only on the antenna's impedance but also the impedance of the receiver itself. One useful result of modeling an antenna as just an impedance value is the simplicity of analysis of the power transfer efficiency from the antenna to the receiver circuits. Consider Figure 5.1a, where the antenna, during receive mode, is modeled as a source of power with a series impedance of 50 Ohms. The receiver is replaced with an impedance value as well. An impedance value is composed of a real (R) part and an imaginary or reactive (X) part. The power transfer efficiency, which is the percentage of the power delivered to the receiver divided by total power in the antenna, can be calculated from the equation:

$$\text{Power Transfer Efficiency} = \frac{2 \times R_A \times R_R}{(R_A + R_R)^2 + (X_A + X_R)^2} \times 100 \qquad (5.11)$$

For example, if both antenna and transceiver have 50 Ohm impedances, the power transfer efficiency will be 50%, which is the maximum theoretical limit. If the receiver impedance

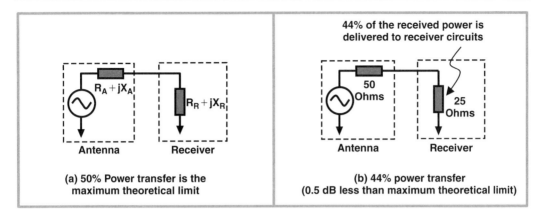

Figure 5.16: Determining the Power Transfer Efficiency Between the Antenna and the Receiver

is reduced to 25 Ohms (Figure 5.16b), the power transfer efficiency will be reduced to 44%. In general, the maximum power transfer efficiency will be achieved when the receiver and the antenna have conjugate matching. Two impedances are conjugate matches if their real parts are the same and their reactive parts have the same absolute value, but with different polarities. For instance, an impedance value of 35 +j28 Ohms is a conjugate match of a 35 − j28 Ohms impedance. If the impedances are not properly matched, additional passive components (resistors, inductors, and capacitors) may be used between the antenna and the receiver to bring antenna and receiver impedances as close as possible to an ideal conjugate match. The spreadsheet provided on this book's companion Website can be used to calculate power transfer efficiency for various scenarios.

5.9.6 Antenna Tuning

Antenna impedance depends on the shape and size of the antenna. *Antenna tuning* is the art of changing antenna impedance by reshaping and resizing the antenna itself. This should not be confused with *antenna matching*, which is the practice of changing the antenna impedance by adding passive components in series and/or parallel of the antenna to match the antenna impedance to the transceiver circuits.

5.9.7 Antenna Polarization

An EM wave consists of an electric and a magnetic field traveling with the same speed (Figure 5.17). The direction of the electric field is defined as the polarity of an EM wave.

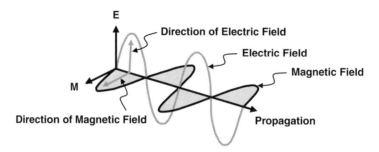

Figure 5.17: An Electromagnetic (EM) Wave Consists of Perpendicular Electric and Magnetic Fields

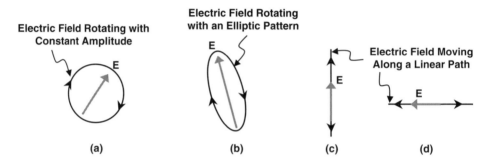

Figure 5.18: (a) Circular, (b) Elliptic, (c) Vertical, and (d) Horizontal Polarizations

There are three polarization types (Figure 5.18): circular, elliptic, and linear (vertical and horizontal). In circular polarization, the phase of the electric field changes with a constant speed and causes the electric field to move along a circle. In elliptic polarization, both amplitude and phase of the electric field change and cause the electric field to rotate in an elliptical form. An EM wave has linear polarization if the electric field moves along a straight line instead of any type of rotation.

Initial polarization of a radio wave is determined by the antenna launching the waves into space. The environment and incidents, such as reflection from a surface, can cause a change in the polarization. The antennas of the receiver and the transmitter nodes, specially in line-of-sight applications, should have the same polarization to maximize the power transfer between the two antennas.

5.9.8 Antenna Options

This section covers the basic characteristics of a number of antennas for ZigBee applications.

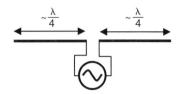

Figure 5.19: Dipole Antenna

5.9.8.1 Dipole Antennas

The *dipole antenna* (Figure 5.19) is one of the simple and widely used antennas in ZigBee applications. This antenna consists of two wires (or traces on a PCB board). The size of each wire is approximately a quarter of the wavelength of the desired frequency of operation. For example, for the 2.4 GHz ISM band with a wavelength of approximately 12 cm, the size of each wire is about 3 cm. Since the total length of this antenna is about half a wavelength, it is also known as a *half-wave* dipole antenna.

Figure 5.20a illustrates the typical radiation pattern for a dipole antenna in two dimensions (*X* and *Y*). The radiation pattern in *X-Z* coordinates is close to an ideal isotropic antenna. The radiation pattern of a dipole in three dimensions (Figure 5.20b) resembles a donut and is often referred to as the *donut-shaped radiation pattern*. The polarization is linear.

An ideal dipole antenna, away from any conducting or dielectric materials, can be approximately modeled by a 73 Ohm resistor. If a dipole is implemented on a PCB, the type of dielectric material and proximity of a ground can change the antenna impedance. The antenna impedance can be tuned by changing the length or shape of the wires. Figure 5.21 shows a number of frequently used shapes of the dipole antenna.

5.9.8.2 Quarter-Wave (Monopole) Antennas

If you remove one half of a dipole antenna and place it perpendicular against a conducting ground plane (Figure 5.22), you have made a *monopole* or *quarter-wave* antenna. The ground plane acts similarly to a mirror and reflects the quarter-wavelength wire. As a result, the propagation characteristic of a quarter-wave antenna is similar to a dipole, above the ground plane. But there is no propagation under the ground plane (Figure 5.23).

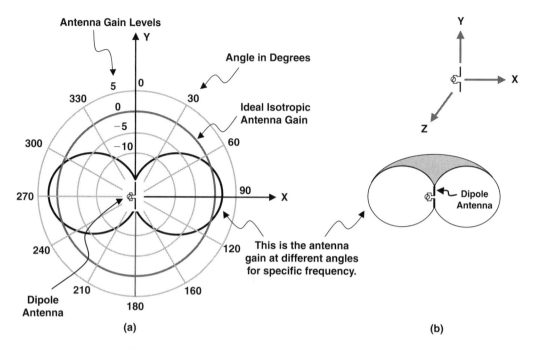

Figure 5.20: Dipole Antenna Radiation Pattern

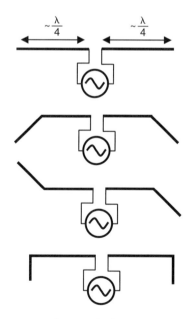

Figure 5.21: Frequently Used Shapes of a Dipole Antenna

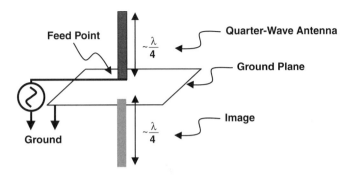

Figure 5.22: Quarter-wave (Monopole) Antenna

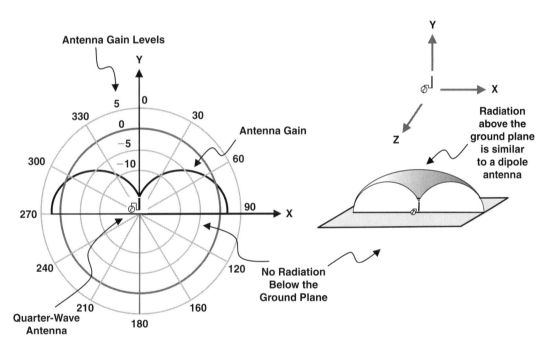

Figure 5.23: Quarter-wave (Monopole) Antenna Radiation Pattern

The impedance of the quarter-wave antenna is 36 Ohms, which is half the impedance of a dipole antenna. Unlike a dipole antenna, a monopole antenna is single ended, which is convenient for interfacing with a single-ended transceiver. This monopole antenna is considered a type of whip antenna because its shape resembles a whip.

5.9.8.3 Tilted Whip or Open-Stub Antennas

The *tilted whip antenna* (Figure 5.24) is a tilted trace on a PCB with a size slightly less than quarter-wavelength. The length of the whip might need to be adjusted depending on PCB thickness and dielectric. The lower the distance between the whip and the ground plane, the lower the antenna impedance and the efficiency. This distance should be at least one-tenth of a wavelength. This antenna has linear polarization and the radiation pattern is close to omnidirectional with a typical gain of around $-10\,\mathrm{dB}i$.

5.9.8.4 Inverted F Antennas

The *inverted F antenna* is one of the commonly used antennas in IEEE 802.15.4 applications because of its simplicity, size, efficiency, and fairly omnidirectional radiation pattern. In an inverted F antenna (Figure 5.25), the impedance is adjusted by changing the location of the feed point. The inverted F antenna impedance can be matched to 50 Ohms.

5.9.8.5 Slot Antennas

A *slot antenna* is made by creating an approximately half-wavelength nonconducting slot in a sheet of metal (Figure 5.26). The propagation characteristic of this antenna is similar to that of a dipole antenna, but the slot antenna impedance is high (several

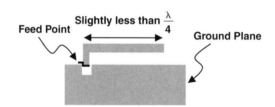

Figure 5.24: Tilted Whip (Open Stub) Antenna

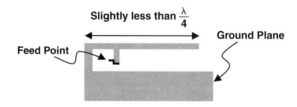

Figure 5.25: Inverted F Antenna

hundred Ohms). The antenna impedance is adjusted by moving the feedpoint location. If this antenna is cut in half, it will become an open-slot antenna with propagation characteristics similar to a monopole antenna.

5.9.8.6 Patch Antennas

Patch antennas come in various shapes and sizes and consist of a patch of metal directly above a ground plane. Figure 5.27 shows an example of a patch antenna. The main disadvantage of these antennas is their relatively large size compared to other types of antennas. For example, some patch antennas are approximately half a wavelength on each side. The polarization can be either circular or linear depending on the design of the patch. In a patch antenna, most of the propagation is above the ground plane and can have high directional gain.

5.9.8.7 Spiral Antennas

If you curl the whip antenna in Figure 5.24 to save area, you have made a *spiral antenna*. The length of the trace is slightly less than a quarter of the wavelength. The spiral antenna has linear polarization. There is no ground plane underneath the spiral. This antenna can become detuned by nearby objects and might not be the best option for handheld wireless applications.

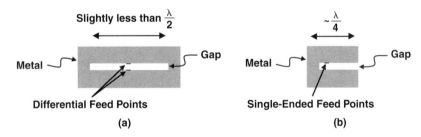

Figure 5.26: (a) Half-wave and (b) Quarter-wave Slot Antennas

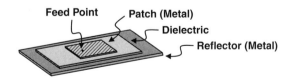

Figure 5.27: Basic Layers in a Patch Antenna

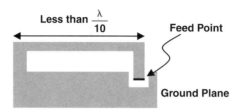

Figure 5.28: Small Loop Antenna

5.9.8.8 Helical (Coil) Antennas

The *helical* (or *helix*) *antenna* is a piece of wire wound into a coil. The coil length is approximately half the wavelength and creates a directional antenna. A helix antenna is bulky and can be detuned by nearby objects. The overall gain of spiral and helix antennas is less than the gain of an inverted F antenna.

5.9.8.9 Chip Antennas

Chip antennas are commercially available surface-mount antennas of reasonable efficiency and small size. A chip antenna can be implemented based on any of the previously discussed antennas, but the footprint is typically smaller than that of the custom antennas. Frequency band of operation is normally narrow and can be detuned easily by nearby objects.

5.9.8.10 Small Loop Antennas

There are two types of loop antenna: *electrically small* loop antennas and *electrically large* loop antennas. If the total loop circumference is less than one-tenth of the wavelength, the loop antenna is considered electrically small. The circumference of an electrically large loop antenna is close to a wavelength. A small loop antenna (Figure 5.28) has narrow frequency bandwidth and requires tuning to ensure the antenna gain is maximized in the frequency band of interest. A small loop antenna has low gain, but it is fairly omnidirectional and is not easily detuned by nearby objects.

5.10 Regulatory Requirements

Regulatory bodies in the United States, Europe, Canada, Japan, and many other nations around the world issue regulations to help maintain wireless communication system integrity and product performance in their regions. The following sections on regulatory are provided for information purposes and the reader is advised to consult the specific regional regulatory document, including the original document and any version(s) superseding it in making a final disposition with regard to any regulatory specifications. In the United States, the Federal Communications Commission (FCC) rules and regulations [10] are codified in Title 47 of the Code of Federal Regulations (CFR). Part 15 of these rules, "Radio Frequency Devices," states that any wireless device sold in the United States is required to receive certification of compliance to part 15. Canadian requirements are usually identical to FCC regulations except for the time lag to align the Canadian rules following FCC rule changes [18]. In Japan, the Association of Radio Industries and Businesses (ARIB) provides the regulatory mandates [12]. In Europe, the Conference of Postal and Telecommunications Administrations (CEPT) provides a high degree of standardization on low-power radio equipment. The European Telecommunications Standards Institute (ETSI) develops standards for CEPT member countries [13].

5.10.1 Brief Overview of FCC Regulations

The operation in 902–928 MHz (a.k.a. the 900 MHz ISM band) and 2400–2483.5 MHz (a.k.a. the 2.4 GHz ISM band) has no limitation for the type of application or transmit duty cycle as long as the signals are digitally modulated and the following two requirements are satisfied:

- Minimum 6 dB bandwidth of 500 KHz
- Maximum spectral density of +8 dBm/3 KHz

The transmit power can be up to 1W (30 dBm), not including the antenna gain. The maximum antenna gain is +6 dBi. If antenna gain is more than +6 dBi, the transmit power should be reduced to keep the overall transmitted power below 36 dBm. The 2.4 GHz ISM band is an exception and the transmit power should be reduced by only 1 dB for every 3 dB that the antenna gain exceeds +6 dBi.

Section 15.205 of the FCC rules specifies certain frequency bands as *restricted bands*, where only a limited amount of spurious emission is allowed. Table 5.5 is a list of these restricted bands above 960 MHz. The second and third harmonics of any signal in the 2.4 GHz ISM band as well as the third, fourth, and fifth harmonics of the signals in the 900 MHz ISM band fall into FCC-restricted bands. The 2.4 GHz ISM band is adjacent to the 2483.5–2500 MHz restricted band, where close-in spurious emissions from the highest frequency channel (centered at 2480 MHz) may easily fall into this restricted band. Some manufacturers avoid using the highest frequency channel of the 2.4 GHz ISM band, to reduce the chance of emission to the adjacent restricted band.

Table 5.5: List of FCC Restricted Bands (in MHz) from 960 MHz to 9.5 GHz

960–1240	1660–1710	2483.5–2500	3345.8–3358	7250–7750
1300–1427	1718.8–1722.2	2690–2900	3600–4400	8025–8500
1435–1626.5	2200–2300	3260–3267	4500–5150	9000–9200
1645.5–1646.5	2310–2390	3332–3339	5350–5460	9300–9500

Microvolts per meter (μV/m) is the unit the FCC uses to specify the maximum allowed strength of an electric field (E) created by a transmitter in the restricted bands. For instance, the amount of emission in any restricted band above 960 MHz should be less than 500 μV/m at 3 m distance. The μV/m unit can be converted to a more well-known unit such as Watt. If the electric field (E) at distance (d) is known, for a unity gain antenna the transmitted power (P) can be calculated from the equation:

$$P = 0.033 \times d^2 \times E^2 \text{ (Watts)} \qquad (5.12)$$

For example, the electric field limit of 500 μV/m at 3 m is equivalent to around 75 nW (-41.2 dBm) transmitted power. Generally, the transmitted power (P) is specified as *effective radiated power* (ERP), which is the power supplied to an antenna multiplied by the antenna gain. For frequencies below 1 GHz, the signal power in the restricted band is measured as the peak power over any 100 KHz frequency band. In the restricted bands above 1 GHz, the signal power is measured as the average signal over any 1 MHz frequency band during worst-case 100 ms periods of operation. The official way of signal averaging and measuring the signal peak is captured in Section 15.35 of the FCC rules.

In any frequency band, the spurious signal power must be at least 20 dB below the maximum signal power in the frequency band of operation when measured over a 100 KHz bandwidth. This is the only spurious emission that needs to be met, if the spurious signals fall outside the restricted bands. But if the spurious signals fall into the restricted bands, the FCC places tighter limitations on spurious emissions. For example, the maximum emission level within the restricted bands above 960 MHz is -41.2 dBm/MHz. A good summary of regulations applicable to IEEE 802.15.4 is provided in the specification [14].

5.10.2   FCC Certification of Compliance

To receive FCC certification of compliance, a product must be sent to an FCC-accredited test lab. The FCC maintains a database of all accredited test labs on its Website. The first step in the certification process is applying for an FCC Grantee Code, which is a code assigned to a specific applicant at a specific address and is the first portion of each FCC Identifier (ID) for devices authorized under the certification procedure. After all tests are completed by the accredited lab, the results will be submitted to the FCC for review. If confidentiality is not requested, all documentation will appear on the FCC Website at the time the certification is granted.

5.10.3 Brief Overview of European Regulations

The wireless devices sold in Europe must receive Conformité Européenne (CE) marking, which is French for European Conformity; they need to comply with Electromagnetic Compatibility (EMC) Directive 89/336/EEC [15]. The European Radio-communications Committee (ERC) Recommendation [16] is a good starting point for understanding European regulations for low-power radio operation.

In European regulatory documents the signals are divided into two categories of narrowband and wideband signals. IEEE 802.15.4 signals are in the wideband category because, per ETSI EN 300 328 document [13], signals modulated via frequency-hopping spread spectrum (FHSS) and direct sequence spread spectrum (DSSS) are considered wideband signals. The emission limits are defined with dBm/Hz units, but actual measurements are taken over a larger bandwidth, such as 100 KHz or 1 MHz. For instance, the -80 dBm/Hz emission limit is equivalent to -30 dBm/100 KHz:

$$-80\,\text{dBm/Hz} + 10 \times \log_{10}(10)^5 = -80\,\text{dBm/Hz} + 50 = -30\,\text{dBm/100 Hz}$$

Maximum in-band transmission power is 100 mW ERP when FHSS is used. For DSSS, the maximum transmit power is 10 mW/MHz. ERC Recommendation 70-30 [16] imposes a maximum active mode duty cycle of 1% (in any one-hour window) for operations in the 868 MHz band. In contrast with the FCC, the ETSI specifies two separate sets of spurious emission requirements for standby and operating modes. European regulations are summarized in Tables 5.6 and 5.7.

5.10.4 CE Conformity Marking

The manufacturer bears the ultimate responsibility for a product's conformity, but the certification process normally involves a combination of self-assessment by the manufacturer and a third-party assessment by a "Notifying Body." The European Union keeps a list of notifying bodies in their New Approach Notified and Designated Organizations (Nando) Information System [17]. This third-party assessment may consist of an audit of the manufacturer's quality system and specific testing of the manufacturer's product.

Table 5.6: Summary of In-Band Transmit Power Limits

Region	Frequency Band	Power Limit	Comment
United States	902–928 MHz 2400–2483.5 MHz	1000 mW/MHz	
Canada	2400–2483.5 MHz	1000 mW/MHz	Some limitations apply based on installation location
Europe	868–868.6 MHz	25 mW	Duty cycle < 1% in any 1h period
	2400–2483.5 MHz	100mW EIRP	For FHSS
		10 mW/MHz	Any other modulation (DSSS, etc.)
Japan	2400–2483.5 MHz	10 mW/MHz	Assuming DSSS

Table 5.7: Summary of Spurious Emission Limits for 30 MHz to 12.75 GHz Bands

Frequency Band	North America	Europe (DSSS or FHSS)	Japan
30–88 MHz	–55.3 dBm/100KHz	–86 dBm/Hz operating –107dBm/Hz standby/receive	–26.02 dBm/MHz
88–216 MHz	–51.8 dBm/100KHz		
216–960 MHz	–49.2 dBm/100KHz		
960–1000 MHz	–41.2 dBm/100KHz		
1000–2387 MHz	–41.2 dBm/MHz	–80 dBm/Hz operating –97 dBm/Hz standby/receive	–16.02dBm/MHz
2387–2400 MHz			
2400–2483.5 MHz	Common ISM Band		
2483.5–2496.5 MHz	–41.2 dBm/MHz	–80 dBm/Hz operating –97 dBm/Hz standby/receive	–16.02dBm/MHz
2496.5–12750 MHz			–26.02dBm/MHz

5.10.5 Brief Overview of Japanese Regulations

In Japan, the 2400–2483.5 MHz band is available for unlicensed applications. The emission requirements are provided in ARIB STD-T66 [17] and are summarized in Tables 5.6 and 5.7. The maximum allowed antenna gain is 2.14 dB compared to +6dB in the FCC rules. Japanese requirements for averaging, peak detection, and spurious emission are not more restrictive than FCC and ETSI rules. Japan also specifies a maximum radiation during receive mode, which is 4 nW for frequencies below 1 GHz and 20 nW for above 1 GHz.

5.10.6 Japan's Conformity Certification System

Any low-power data communication system sold in Japan is required to receive ARIB conformity certification. The Telecom Engineering Center (TELEC) is the official radio laboratory for the Japanese government radio regulatory office [19]. TELEC was chartered by the Japanese Ministry of Posts and Telecommunications and has been made a Designated Certification Agency.

The in-band transmit-power limits and maximum out-of-band emission requirements for North America, Europe, and Japan are summarized in Tables 5.6 and 5.7.

References

[1] S. Y. Seidel and T. S. Rappaport, "914 MHz Path-Loss Prediction Models for Indoor Wireless Communications in Multifloored Buildings," *IEEE Transactions on Antenna and Propagation,* Feb. 1, 1992, pp. 207–217.

[2] T. S. Rappaport, *"Wireless Communications: Principles and Practice,"* Prentice Hall, NJ, 2002.

[3] D. Kim, M. Ingram and W. Smith, "Measurements of Small-Scale Fading and Path Loss for Long-Range RF Tags," *IEEE Transactions on Antennas and Propagations.,* Vol.51, Aug. 2003, pp. 1740–1749.

[4] ITU Recommendations, "Propagation Data and Prediction Methods for the Planning of Indoor Radio-Communication Systems and Radio Local Area Networks in Frequency Range 900 MHz to 100 GHz," ITU-R 2005.

[5] H. L. Bertoni, *"Radio Propagation for Modern Wireless Systems,"* Prentice Hall, 2000.

[6] J. S. Seybold, *"Introduction to RF Propagation,"* John Wiley & Sons, 2005.

[7] W. C. Y. Lee, *"Mobile Communication Engineering, Theory and Applications,"* 2nd ed., McGraw-Hill, NY, 1998.

[8] Y. L. C. Jong and M. Herben, "A Tree-Scattering Model for Improved Propagation Prediction in Urban Microcells," *IEEE Transactions on Vehicular Technology,* March 2004, pp. 503–513.

[9] J. Medbo and P. Schramm. "Channel Models for HIPERLAN 2. ETSVBRAN," document no. 3ERI085B, 1998.

[10] Federal Communications Commission (FCC), available at www.FCC.gov.

[11] H. Sizun, *"Radio Wave Propagation for Telecommunication Applications,"* Springer, 2004.

[12] Association of Radio Industries and Businesses (ARIB), available at www.arib.or.jp.

[13] European Telecommunications Standards Institute (ETSI), available at www.etsi.org.

[14] "IEEE 802.15.4: Wireless Medium Access Control (MAC) and Physical Layer (PHY) Specifications for Low-Rate Wireless Personal Area Networks (WPANs)," Sept. 2006.

[15] EMC Directive 89/336/EEC, available at http://ec.europa.eu/.

[16] ERC Recommendation 70-03, available at www.ero.dk.

[17] Nando Information System, available at http://ec.europa.eu/enterprise/newapproach/nando.

[18] Strategis, Produced by Industry Canada, available at http://strategis.ic.gc.ca.

[19] Telecom Engineering Center (TELEC), available at www.telec.or.jp.

[20] D. M. Dobkin, *"RF Engineering for Wireless Networks"*, Elsevier Science, 2005.

CHAPTER 6

Battery Life Analysis

This chapter reviews the battery discharge characteristics, provides a simple method of battery life calculation, and discusses some of the power reduction methods used to prolong network lifetime. Battery life is a function of both hardware-level performance and network operation efficiency.

6.1   Battery Discharge Characteristics

Battery capacity is measured in milliamps $\times$ hours (mAH). For example, if a battery has 250 mAH capacity and provides 2 mA average current to a load, in theory, the battery will last 125 hours. In reality, however, the way the battery is discharged has an impact on the actual battery life. Discharging a battery at the manufacturer-recommended rate normally helps the battery deliver close to its nominal capacity. But the result cannot simply be extrapolated linearly to other discharge profiles.

In many short-range wireless sensor networks, although the average current consumption of a device is low, the instantaneous current can be high. For example, in Figure 6.1, a transceiver with sleep current of 1 μA and peak active current of 20 mA wakes up every 2 seconds and stays active for 5 ms. The average current consumption is only 50 μA, but the peak current of 20 mA can have an adverse effect on the battery's actual capacity, especially if there is not enough time between the periods of high current discharge to let the battery rest and recover.

This can be explained by the relaxation phenomena (or recovery effect) [1]. When a battery is discharged at a high and sustained rate, the battery reaches its end of life even if there are still active materials left in the battery. However, if the discharge rate is not continuous and there are cutoffs or very low-current periods, the transport rate of active

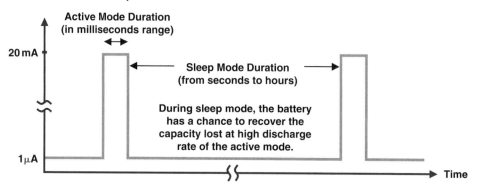

Figure 6.1: An Example of Current Profile of a Device in a ZigBee Network

materials catches up with the depletion of the materials, giving the battery a chance to recover the capacity lost at the high discharge rate.

The actual capacity of a battery for a specific use-case scenario can be determined experimentally. Battery manufacturers might be able to provide an estimate for the actual battery capacity if the application's current profile (e.g., Figure 6.1) is provided. The ratio of the battery's actual capacity to its nominal capacity is called the *battery efficiency factor*.

One way to avoid high current discharge periods, if possible, is to use a sufficiently large capacitor to supply the current to the transceiver when the node is active. While the device is in sleep, the battery charges the capacitor, and when the device becomes active and requires a high discharge rate, this capacitor will provide current to the device. In this way, the battery would not experience high discharge rate periods and the battery efficiency can be improved.

Another characteristic of a battery is its self-discharge rate. A battery, even when not in use, loses its capacity over time due to internal leakage. Battery manufacturers quantify this leakage as *self-discharge per month*. For example, a 300 mAH battery with self-discharge per month of 0.5% loses 1.5 mAH of its capacity after one month. The battery shelf life is defined as the longest time a battery can be stored before its capacity falls below 80% of its nominal.

6.2   A Simple Battery Life Calculation Method

The first step in battery life calculation is simplifying the use-case scenario. An example is shown in Table 6.1. In this example, the device wakes up every hour by its internal clock, gathers sensor information, performs CCA, and transmits the data to a

Table 6.1: A Simplified Use-Case Scenario

Step	Action	Duration	Average Current	Energy (mAH)
1	Device is in sleep	1 hour	1 μA	1.00×10^{-3}
2	Device transitions to active mode	10 ms	50 μA	1.39×10^{-7}
3	Sensor information is captured, processed, and stored	1 ms	5 mA	1.39×10^{-6}
4	Device is in receive mode to perform CCA	700 μs	20 mA	3.89×10^{-6}
5	Device transmits the packets	550 μs	20 mA	3.06×10^{-6}
6	Device is in receive mode waiting for an acknowledgment	400 μs	20 mA	2.22×10^{-6}
7	Device goes back to sleep			
Total Energy for this event (mAH):				1.01×10^{-3}

coordinator. Finally, the device goes back to sleep after receiving the acknowledgment. If the channel is not clear or the acknowledgment is not received, the device repeats steps 4–6 a predetermined number of times before it declares the transmission unsuccessful. The battery capacity used in this case can be easily calculated by multiplying the time duration of each step by its associated current consumption.

An interesting observation from Table 6.1 is the fact that 1×10^{-3} mAH is used during sleep mode compared to the total active mode 1×10^{-5} mAH usage. It means that in extremely low-duty-cycle applications, the sleep mode current can have a significantly higher impact on determining battery life than the active mode average current.

The second step is determining the battery efficiency for the specific application scenario. This can be done by either experiment or inquiring of the battery manufacturer. For example, if battery nominal capacity is 300 mAH with efficiency of 60% for a specific application, 180 mAH is the actual capacity that should be used in battery life calculations.

The self-discharge of the battery can be an important factor in battery life in very low-duty-cycle applications. Consider a battery with actual capacity of 500 mAH and a self-discharge rate of 1% per month. In the application scenario of Table 6.1, after one month, the total capacity used by the device during active mode and sleep mode is 0.009 mAH and 0.72 mAH respectively. In comparison, the energy wasted due to the battery self-discharge over one month is 5 mAH, which therefore has the biggest impact on the battery life.

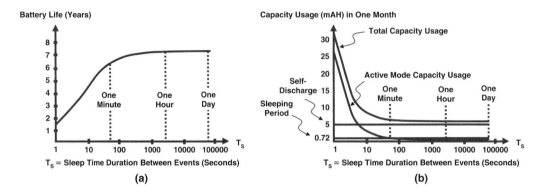

Figure 6.2: (a) Battery Life for the Scenario Shown in Table 6.1. The Battery Has 1000 mAH Nominal Capacity, 50% Battery Efficiency, and 1% Self-discharge Per Month. (b) Total Capacity Used in One Month for Different Activities

The simplest way to approximate battery life is to divide the battery's actual capacity by the total capacity used by the device and drained due to self-discharge. On this book's companion Website you will find a simple battery life calculation spreadsheet that can be used for various use-case scenarios.

Figure 6.2a shows the effect of duty cycle on battery life. As the sleep duration between events increases, battery life improves. But improvement slows down at higher time intervals because the active current is no longer a major contributor to battery life. Figure 6.2b represents the total capacity used during active mode, sleep mode, and battery self-discharge over a one-month period. For example, if the sleep duration between events is one second, the total energy consumed during active mode over one month is 27 mAH. However, if the sleep duration is increased to one minute, the device wakes up less frequently and the total energy consumed during active mode over one month becomes only 0.46 mAH. This graph indicates that the active mode current consumption becomes a negligible contributor for sleep durations above one hour.

6.3 Battery Monitoring

In most batteries, the battery's internal resistor increases as the remaining capacity of the battery is reduced. The higher the internal resistor, the lower the battery voltage. During active mode, when the highest current level is drawn from the battery, this voltage drop will be at its maximum. For example, a fresh AA battery may not fall below 1.5 V while providing 20 mA but could drop to 1.0 V as it reaches its end of life.

Therefore, the battery voltage, while being discharged at high current, can be a simple indicator of the battery's remaining capacity. If the transceiver is equipped with an ADC, one of the easiest ways to monitor battery level and generate a low-battery warning when necessary is using one of the transceiver ADC channels to check the battery voltage on a periodic basis. The ADC should read the battery level when the device is active and battery discharge rate is high.

Knowing the battery level will help you plan ahead and potentially make some changes in the way a device is used so that you can extend its remaining life.

6.4 Power Reduction Methods

There are two ways to reduce the power consumption in a wireless network: selecting better hardware and improving the network operating efficiency. To improve battery life, it is important to explore both methods. The following subsections review some of the recommended techniques in each approach.

6.4.1 Hardware-Level Considerations

Power reduction at the hardware level is intuitive. The transceiver active current as well as the current consumption during sleep mode directly impact battery life. *Dynamic power management* (DPM) allows a device to have several modes of operation at the hardware level. The DPM provides the option to turn on only the subcircuits required to accomplish a task and conserve the battery energy that otherwise would be consumed by unnecessary subcircuits. For example, when reading sensor information using an ADC, it is not necessary to have radio circuits up and running until after the data is ready to be transmitted.

Table 6.2 shows a simplified example of the modes of operation in a wireless device. The ON mark in the table means that a specific block is turned on. Mode 1 consumes the least amount of current because there no is active block in this mode and the only current consumed is the leakage current from the power supply to the device. There is no clock running in mode 1; therefore, the only way to come out of mode 1 is via an external trigger. In mode 2, only the internal low-accuracy clock and its regulator, if present, are running, which allows the device to keep track of approximate time while maintaining very low current consumption (e.g., less than $1\,\mu\text{A}$ total). Mode 3 keeps track of time more accurately using a crystal oscillator at the cost of higher current consumption. Mode 4, for example, can be used to gather sensor information and store it in memory. Mode 5 is normally a transition mode between receiving (mode 6) and transmitting (mode 7).

Table 6.2: Simplified Example of Modes of Operation

Mode	Receiver	Transmitter	Regulator(s)	High-Frequency Crystal Oscillator	Low-Accuracy Oscillator	RAM	MCU	ADC
Mode 1								
Mode 2			ON		ON			
Mode 3			ON	ON				
Mode 4			ON	ON		ON	ON	ON
Mode 5			ON	ON		ON		
Mode 6	ON		ON	ON		ON		
Mode 7		ON	ON	ON		ON		

In simple applications, the microcontroller unit (MCU) power consumption may have a small impact on battery life. But if the device uses the MCU to run computationally intensive algorithms, the MCU can become a major source of power consumption. The power consumed by an MCU is proportional to both the MCU core voltage (V) and the clock frequency (f):

$$\text{MCU power consumption} \propto fV^2 \qquad (6.1)$$

Reducing the clock frequency or core voltage will reduce the instantaneous power consumption. The only drawback associated with lower core voltage is that the MCU cannot operate at its maximum clock frequency. This is due to the fact that CMOS logic, used in the majority of microprocessors, has a voltage-dependent maximum operating frequency, and when the core voltage is reduced, it is necessary to reduce the clock frequency as well. Lowering the clock frequency will increase the delay in the system because the device needs more time to complete its computational duties. Normally, in low-duty-cycle and low-data-rate wireless sensor networking, this delay is not expected to cause any issue. This method is referred to as *dynamic voltage scaling* (DVS) because the device can dynamically change the MCU power consumption and clock frequency based on the required performance in various scenarios. This technique is also known as *dynamic voltage and frequency scaling* (DVFS).

If the application is not time-critical, it is possible to use a low frequency and a low-power clock with degraded accuracy instead of a precise clock that typically consumes

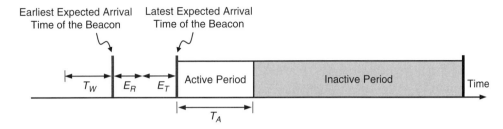

Figure 6.3: A Node Must Wake Up Sufficiently Earlier Than the Beacon-expected Time of Arrival

much higher power than a low-accuracy clock. The low-accuracy clock is not intended for beacon-enabled operation where the device is expected to wake up just before the beacon arrives. The clock inaccuracy causes an uncertainty in expected arrival time of the beacon frame. Figure 6.3 shows the earliest and latest expected beacon arrival time due to timing errors in receiver and transmitter. The total timing error associated with the clock inaccuracy in the receiver is shown by E_R. The value of E_R is a function of beacon interval. For example, if the receiver crystal frequency has $+/-40$ pm of error and the beacon interval is 61.44 ms, the receiver timer will have $+/-2.46\,\mu s$ of error after 61.44 ms is elapsed ($E_R = 2.46\,\mu s$). This means that the timer accumulates the error over time, and the longer the beacon interval, the larger the accumulated error.

The timer error in the beacon transmitter (E_T) is calculated similarly. To make sure the receiver is ready when the beacon arrives, the receiver must wake up $E_R + E_T$ seconds before the expected beacon arrival time. If the receiver goes into sleep mode during the inactive period, the receiver will require a warmup period (T_W) prior to the expected beacon arrival. Therefore, in a beacon-enabled network, a receiver needs to stay active for $T_W + E_R + E_T$ seconds in addition to the active period (T_A) after the beacon frame. The clock frequency inaccuracy of the low-power/low-accuracy clock in mode 2 of Table 6.2 can be as high as 30% and significantly increase the $E_R + E_T$ period. That is the reason a low-power/low-accuracy clock is not recommended in a beacon-enabled operation.

A node may spend the majority of its time in sleep mode but can wake up very frequently, depending on the application scenario. The current consumed during the warmup period can be low, but if the circuit startup time is very long, the energy consumed during the wakeup period can noticeably degrade the device's battery life.

If a node is placed in a location in which all the communication with other nodes is always performed in a specific direction, a high-gain directional antenna can be a better choice than an omnidirectional antenna because the additional gain available in one direction allows the transmitter to reduce its current consumption while delivering the same RF output power in the desired direction. This is due to the fact that the energy consumed in a power amplifier (PA), which is the last stage of a transmitter, scales with the amount of RF output power it delivers to the antenna. If an antenna has positive gain, the required RF output power from the PA is reduced and the PA can scale down its output power, which reduces the energy consumed by the PA. The directional antenna also improves the sensitivity level, which increases the chance of successful packet reception.

Energy scavenging (or harvesting) is the ability to harvest a small amount of energy from sources such as heat, light, or motion. The availability of a secondary source in addition to the battery can help a device prolong its battery life. Energy harvesting may not be a very low-cost solution, and the source of the energy might not be present all the time. The amount of energy received from the environment can vary over time. Depending on the application, the energy harvesting might not be reliable enough to be the only source of the energy on a node.

6.4.2   Network Operation Efficiency

The previous section discussed the relationship between the hardware-level performance of a node and battery life. The maximum current consumption during active mode and the amount of leakage current during sleep mode are examples of hardware-level performances that directly affect battery life.

Battery life not only is a function of hardware-level performance of each node but also depends on the operation efficiency of the network. For example, simplifying the networking protocol itself can reduce the number of operations a device needs to perform in establishing communication links with other nodes in the network. The routing mechanism and selection of the link cost can greatly impact network operation efficiency. The ZigBee standard allows the network installers to select any link-cost calculation method they find most suitable for their application. A number of *efficient routing methods* are discussed in Section 6.4.3.

When a device tries to transmit data to another device, the total energy required for successful transmission of the data must be taken into account instead of instantaneous power consumption. The *goodput* is defined as the number of useful bits transmitted per second, excluding the packet overhead and retransmissions. Any packet in IEEE 802.15.4 starts with a preamble field. Recalling from Table 3.4 in Chapter 3, the duration of the preamble in each packet can vary from 120 microseconds to 1600 microseconds.

Considering that the packet payload may be as short as several hundred microseconds, the time dedicated to transmitting and receiving the preamble can be a significant portion of the total time a device spends in active mode. The energy per goodput bit, in Joules per bit (*Jpb*), can be used as a figure of merit to compare the performance of various networking methods:

$$\text{Energy per goodput bit } (Jpb) = \frac{\text{Averaged Consumed Power } (W)}{\text{Goodput } (bps)} \qquad (6.2)$$

If a device needs multiple attempts to successfully transmit a packet, the goodput stays the same, but the energy per goodput bit will increase almost proportionally to the number of attempts.

It is important to remember that ZigBee and other standards developed for low-data-rate applications might have higher energy per goodput bit compared to high-data-rate standards (e.g., IEEE 802.11). This is due to the fact that in low-data-rate applications, the energy consumed during active mode is generally less critical than the energy dissipation during sleep mode, and the extended battery life in low-data-rate applications is achieved by lowering the duty cycle and not by reducing the energy per goodput bit. Therefore, Equation 6.2 should only be used to compare the standards developed for the same type of applications.

Unless the receiver is located in a multipath null, increasing the RF transmission power generally results in improvement in the signal-to-noise ratio of the signal at the receiver and consequently improves the packet error rate (PER). One power-saving method is to adjust the RF transmission power according to the packet size. For a given bit error rate (BER), a longer packet has a higher chance of error compared to a shorter packet. Increasing the RF transmission power proportionally to packet length can potentially improve the PER and reduce the number of retries required to successfully deliver the packet to the destination.

If the physical locations of the devices are known, the RF transmission power can be adjusted to optimize the battery life based on the physical distance between the transmitter and receiver. However, the channel characteristics change dynamically and the RF output setting might need periodic adjustment. Reducing the RF output power will reduce the instantaneous power consumption of the device, but if the power setting is too low for a given environment, the packets might need to be transmitted several times before an acknowledgment is received.

In a fading environment, if the device is located in a multipath null, increasing the RF transmission power might not make any difference. Antenna diversity can improve the chance of successful packet delivery in a fading environment and reduce the number of attempts required to deliver a packet.

IEEE 802.15.4 uses CSMA/CA for channel access. Every time the channel is busy, the device needs to back off for a random period of time. As discussed in Chapter 3, the battery life extension (BLE) option allows a coordinator to turn off its receiver after a period equal to *macBattLifeExtPeriods* following the transmission of a beacon frame to conserve energy. It means that the back-off period is limited to the lesser of 2 and the value of *macMinBE*.

Another source of battery life degradation in a network is the hidden and exposed nodes issues discussed in Chapter 3. In the hidden node issue, the collision of packets results in unsuccessful packet delivery and the packets must be retransmitted. In the exposed node condition, the unnecessary prevention of a node transmitting a packet forces the device to consume battery energy to repeat the CCA, whereas the node could have started its transmission after the first CCA. Changing the location of the nodes, adjusting the RF output power of the nodes, and using directional antennas are examples of the methods that can be used to overcome the hidden and exposed node problems.

6.4.3   Energy-Efficient Routing

The definition of network lifetime is application specific. The lifetime of a network is often defined as the duration of the time from the network is established until the first node fails due to battery depletion. An alternative way to define network lifetime is the maximum number of times a certain task (e.g., data collection) can be carried out without any node running out of energy. If the overall remaining energy level in the network is more critical than the energy exhaustion in a single node, the lifetime can be defined as the summation of remaining energy of the individual nodes in the network. The definition of network lifetime determines the method we should use to improve energy efficiency. For instance, if only the overall energy consumption matters, a routing method that minimizes the total energy consumed to route each packet would be the optimum choice. This approach, however, might not be appropriate if the network lifetime is defined as the duration until the first node fails. The reason is explained in Section 6.4.3.1.

If all the routers in a ZigBee network are connected to a main supply, the energy efficiency in routing is not critical and the implementer can focus on other performance metrics, such as reliability and latency. However, if the majority of the routers are battery powered, the routing energy efficiency becomes an important metric. Selecting proper routes can prolong the network lifetime. In implementing a ZigBee network, we are bound by the route discovery procedure in the ZigBee NWK layer. But the application developer can influence the routing procedure, for example, by defining the link cost. ZigBee allows the user to have flexibility in selecting the link cost based on the application scenario. The link cost in route discovery can be a function of the available information such as the link quality, the distance between the nodes, or the status of the battery in each node.

There is no universally optimum routing mechanism and the best routing method may vary based on the use-case scenario. Furthermore, in an event-driven environment, there is no prior knowledge of the future packet arrivals in each node, which adds to the challenge of optimizing routing efficiency. Therefore, the route selection mechanism may need to be modified over time based on actual packet traffic distribution in the network. In the following subsections, a number of methods used to improve routing efficiency are reviewed. Generally, the goal of routing efficacy is to prolong the network lifetime while avoiding poor link quality or very long message latency. Poor link quality not only can be unacceptable for the application, but also multiple attempts to deliver a message can degrade the network lifetime.

6.4.3.1 *Battery-Aware Routing*

The basic route discovery method in ZigBee uses LQI to generate the link-cost function. Since this simple method does not include energy consumption as part of the link cost, this approach is not considered an energy-aware mechanism. Energy awareness can be added to this routing method if the routes are selected based on a trade-off between the overall LQI and the total energy consumed to route a packet. This modification improves the energy consumption in each individual packet delivery but can have one shortfall. Depending on network configuration, it is possible that some of the nodes participate in routing very often, whereas many other nodes hardly relay a packet. As a result, some of the nodes will drain their batteries very quickly. This issue can be resolved by shaping the packet traffic based on battery status of the nodes in the network.

The node power descriptor contains the information regarding the battery status (Table 3.13). The remaining (residual) energy level of a node battery can be taken into account when calculating the link cost. For example, the link cost can be a function of the LQI as well as the residual energy of the battery. By increasing the link cost to the nodes with low residual energy, the new routes will pass through these nodes less often. This will help shape the packet traffic away from the nodes with critically low battery levels as much as possible. This packet traffic shaping can help the nodes in a network drain their batteries more uniformly.

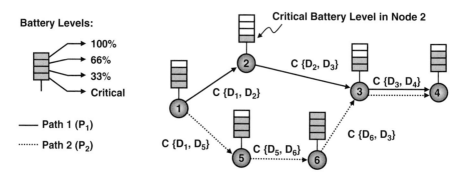

Figure 6.4: Residual Energy-Aware Routing

Figure 6.4 shows an example of residual energy-aware routing. Path 1 has lower latency and requires less overall energy to relay packets from node 1 to node 4. But node 2 has critically low residual energy left. Therefore, $C\{D_1, D_2\}$ is increased high enough to ensure that the packets are routed through path 2 as long as the routing tables of nodes 5 and 6 are not full and the overall LQI and latency are acceptable.

If a device is capable of energy scavenging, the information regarding the availability of harvested energy can be taken into account in calculating the cost to a node in addition to the residual energy of the battery. The link cost to a node with energy-harvesting capability can be lower than to a node with similar LQI and residual energy.

The node power descriptor also specifies the type of power source in a node. If a node has a main supply, the residual energy-aware routing will try to route the packets through the main supply-powered nodes as much as possible. The link cost to a main powered node is less than any other node with similar LQI level.

Battery failure can cause a network to become partitioned. Figure 6.5a shows that if the residual energy levels of the nodes are not taken into account in selecting the routes, the nodes at the middle can be used more often and their batteries can be drained completely. As a result, two sections of the network cannot communicate with each other, and one part of the network loses the connection to the PAN coordinator. In Figure 6.5b, residual energy-aware routing resulted in relatively uniform battery consumption in the network and the network integrity is still in place.

The low duty cycle of most wireless sensor networking applications provides the opportunity of enhancing the battery capacity based on the relaxation phenomenon discussed in Section 6.1. During the sleep (idle) mode, the battery can partially recover the capacity lost during active mode. This information can also be taken into account

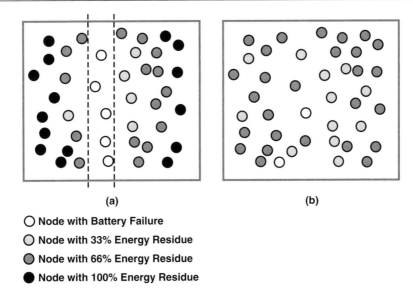

(a)   (b)

○ Node with Battery Failure
○ Node with 33% Energy Residue
◑ Node with 66% Energy Residue
● Node with 100% Energy Residue

Figure 6.5: (a) A Network Partitioned Due to Battery Failure and (b) An Example of Relatively Uniform Energy Consumption Across the Network

in determining the link cost. In this method, the battery status depends not only on the remaining energy but also on the transmission rate of the node. A higher transmission rate means that the battery has less time to recover between transmissions. The link cost to a node with a higher transmission rate can be more than a node with similar residual energy but lower transmission rate.

6.4.3.2 Location Aware Routing

The physical location information, if present, can be used to as part of route discovery to improve the overall routing efficiency and control the traffic flow. The approximate physical location of the nodes in a short-range wireless network can be determined using the algorithms discussed in Chapter 7. Many sensor networking and monitoring applications may require knowledge of the approximate location of the nodes regardless of the routing mechanism.

Knowledge of the location of the next hop not only helps a node adjust the RF output power accordingly, it also determines the amount of progress made toward the destination if the node is selected as the next hop. With this information, along with the LQI, residual energy in the battery can be taken into account in determining the link cost. Figure 6.6 shows a scenario in which a packet is sent from node 1 to node 5. In each hop, the packet

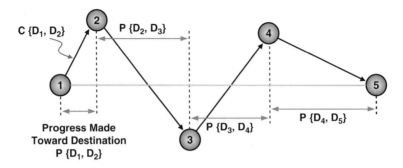

Figure 6.6: Progress Made Toward the Destination

makes some progress toward or away from the destination. For example, the progress made toward the destination by sending the packet from node 1 to node 2 is shown by $P\{D_1, D_2\}$. If the progress is normalized to the total linear distance from node 1 to node 5, the summation of all the progress will be equal to 1:

$$P\{D_1, D_2\} + P\{D_2, D_3\} + P\{D_3, D_4\} + P\{D_4, D_5\} = 1 \qquad (6.3)$$

The cost function $C\{D_1, D_2\}$ can be selected to be inversely related to $P\{D_1, D_2\}$. In this way, the more progress made toward the destination by selecting a node as the next hop, the lower the link cost will be. If the next hop takes the packet away from the destination, the progress function P will be negative.

To take energy efficiency into account, the link cost will not only be a function of progress toward destination and link quality, but also the information such as the residual energy of the batteries of the next-hop neighbors and the energy consumption required to relay the packet to each neighbor can be taken into account. Each input will have a different weight in determining the overall link cost based on the application scenario.

If the location information is known, the flow of the packets in a network can be managed by reducing the link cost when the next hop is comparatively in line with the desired flow. Shaping the traffic can help distribute the energy consumption more evenly and improve the network lifetime.

6.4.3.3 Increasing Routing Efficiency Using a Directional Antenna

If in addition to knowledge of the location of other nodes, each node is equipped with a steerable directional antenna, the energy efficiency of the routing can be further improved. A *directional antenna* is an antenna that has high gain toward one direction. The direction of an electronically steerable antenna can be changed without any mechanical movement. Mechanical movements can consume large quantities of energy and therefore might not be suitable for low-power wireless sensor networking.

Electronically steerable directional antennas can considerably increase the cost and size of a node compared to using a basic antenna with fixed propagation characteristics. One simple way of implementing an electronically steerable directional antenna is to place multiple directional antennas (with fixed directions) and point them to different directions. Then connect one antenna at a time to the radio and disconnect the rest.

The antenna gain helps reduce energy consumption when the node actively receives or transmit packets. In addition, undesired interferences are reduced compared to using an omnidirectional antenna because the transmitted signal is directed toward its intended recipient. Furthermore, if the directional antenna has a narrow coverage, the frequency channel the node uses will be available in the areas of the network not covered by the directional antenna. For example, in Figure 6.7, node A is using a 2450 MHz channel to transmit signals toward the shaded region. But the same frequency channel is available to use in other regions as long as the coverage of the directional antenna of the nodes transmitting concurrently does not overlap. These properties of directional antennas can effectively reduce the number of attempts required to successfully transmit a packet and, consequently, prolong the network lifetime.

The drawback of this method is additional complexity. Not only does each node need to know the physical location of other nodes, but also when steerable directional antennas are used, the node needs to point its antenna in the right direction at the right time.

When the antenna is directional, the *route request* command is sent in one direction (in most cases, toward the destination). In this way, some of the next-hop neighbors have a

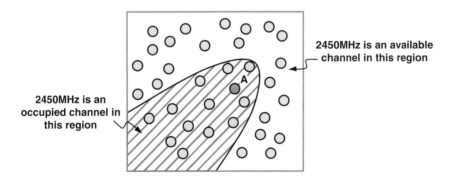

Figure 6.7: A Directional Antenna can Reduce Unnecessary Interference in Some Regions of the Network

higher chance of receiving the *route request* command with better link quality. In other words, the search for the next hop is performed in a limited section of the network. In contrast, an omnidirectional antenna broadcasts the route request in all directions and may unnecessarily engage many nodes in route discovery. A node with a steerable antenna has the option to search for a proper next hop in other directions if the search in one direction is not successful.

6.5 Buck Converters

A typical coin cell battery or two AAA batteries in series deliver 3.0 V to 3.6 V nominal voltage. Most transceivers, however, are satisfied with a smaller voltage (e.g., 2.0 V or less) on their supply pin to operate properly. If the supply voltage is higher than the minimum required voltage, the transceiver internal circuitries (regulator blocks) will regulate down the external supply voltage to their desired voltages.

In a typical transceiver, the current consumption may be independent of battery voltage. It means that providing additional voltage beyond the minimum required for proper operation of the transceiver does not reduce the amount of current extracted from the battery. For example, if the minimum supply voltage for a transceiver is 2.0 V but the series AA batteries provide 3.3 V, the transceiver might not take advantage of the additional 1.3 V. However, if a buck converter is present, it is possible to use this additional voltage to reduce the transceiver's average current consumption.

Figure 6.8a shows the basic concept, where the battery voltage is converted down to a minimum of 2 V. Instead of continuously drawing current from the battery, a switch

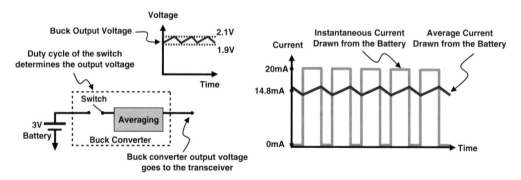

Figure 6.8: (a) Basic Concept of Buck Converter Operation Described in an Example where a 3 V Supply is Converted to 2 V by Means of Switching and Averaging. (b) Instantaneous Versus Average Current Drawn from the Battery

periodically connects and disconnects the battery to an averaging circuit. The duty cycle of this switch will determine the output voltage. In this example, the output of the averaging circuit is a continuous voltage with minimum of 2 V. Considering that the battery is not continuously discharged, the average current drawn from the battery is reduced. The combination of the switching circuit and averaging is known as a *buck regulator* or *DC-to-DC converter*. The efficiency of converting a DC voltage to a lower-value DC voltage by means of switching and averaging is called *buck converter efficiency*, which can be as high as 90% or more.

A rough estimate for the average current saved by using a buck converter is provided in the equation:

$$I_B = I_T \frac{\text{Buck converter average output voltage}}{\text{Battery voltage} \times \text{Converter efficiency}} \tag{6.4}$$

where I_B is the average current drawn from the battery and I_T is the average current delivered to the transceiver. For example, if a buck converter with 90% efficiency converts a 3 V battery voltage to 2 V supply while delivering 20 mA to the transceiver, then the average current drawn from the battery will be 14.8 mA instead of 20 mA.

A buck converter may introduce additional undesired spurious on the supply voltage, which can have adverse effects on sensitive RF blocks. Furthermore, if the buck converter is not designed properly, it will have periods of very high current extraction, referred to as *in-rush current*, resulting in high discharge rate of the battery. These high-current periods are very short, but they can adversely affect the battery capacity, as discussed in Section 6.1.

References

[1] T. F. Fuller, M. Doyle and J. Newman, "Relaxation Phenomena in Lithium-Ion Insertion Cells," *J. Electrochem. Soc.*, Vol. 141, No. 4, April 1994, pp. 982–990.

[2] H. Gossain, et al. "DRP: An Efficient Directional Routing Protocol for Mobile Ad Hoc Networks," *IEEE Transactions on Parallel and Distributed Systems*, Vol.17, Dec. 2006, pp. 1438–1541.

[3] "IEEE 802.15.4: Wireless Medium Access Control (MAC) and Physical layer (PHY) Specifications for Low-Rate Wireless Personal Area Networks (WPANs)," Sept. 2006.

[4] A. Spyropoulos and C. S. Raghavendra, "Energy-Efficient Communications in Ad Hoc Networks Using Directional Antennas," *Proceedings of IEEE INFOCOM,* Los Angeles, 2002, pp. 220–228.

[5] J.-P. Ebert, "Energy-Efficient Communication in Ad Hoc Wireless Local Area Networks," Technische Universität Berlin, Dissertation, April 2004.

[6] K. Wang, J. G. Proakis and R. R. Rao, "Energy-Efficient Routing Algorithms Using Directional Antennas for Mobile Ad Hoc Networks", *International Journal of Wireless Information Network,* Springer Netherlands, April 2002, pp. 105–118.

[7] K. Zeng et al., "Energy-Aware Geographic Routing in Lossy Wireless Sensor Networks with Environmental Energy Supply," 3rd ACM International Conference on Quality of Service in Heterogeneous Wired/Wireless Networks, Waterloo, Canada, Aug. 2006, Vol. 191, Article No. 8.

CHAPTER 7

Location Estimation Methods

This chapter reviews three methods that can be used in an IEEE 802.15.4 network to determine the location of an object. The first one uses received signal strength (RSS) as a simple way of estimating the distance between nodes. The second approach takes advantage of the signal angle of arrival, if known, at two or more nodes to estimate location of the node that transmitted the signal. The last method measures the time difference of signal arrival at multiple nodes with known locations to estimate the location of the node of interest. Practical limitations of these approaches are briefly reviewed here. Among these three methods, the RSS-based location estimation has received the most attention because of its minimum hardware requirements and the simplicity of its implementation.

7.1   Introduction

One of the applications of short-range wireless networking is determining the approximate physical location of objects at any given time. The real-time knowledge of the location of personnel, assets, and portable instruments can increase management efficiency. *Location estimation* refers to the process of obtaining location information on a node with respect to a set of known reference positions. The location estimation is also referred to as *positioning*, *locationing*, and *geolocationing*. The knowledge of the location of the nodes presents the opportunity of providing location-dependent services. For example, a visitor in a museum can carry an audio/video device that provides relevant information to the visitor, depending on his or her location in the museum. The location of a node also can be used as part of the authentication process. In this way, the authenticity of a packet is determined not only by the information embedded in the packet but also by the location of the node that transmitted the packet.

This chapter focuses only on the location-estimation methods that use short-range radio frequency (RF) signals. However, it is possible to use other types of signals such as ultrasound or infrared instead of an RF signal in a location-estimation algorithm. Ultrasound is a low-frequency signal (e.g., 40 KHz) that is above human hearing capability. An ultrasound signal travels with the speed of sound and if its travel time from the transmitter to the receiver is known, the approximate distance between the transmitter and receiver can be determined by multiplying the travel time by the speed of sound. An infrared signal travels with the speed of light. But in contrast to an RF signal, an infrared signal cannot penetrate objects. The systems implemented based on ultrasound or infrared are normally dedicated only to determining the estimated location of the objects. In an RF-based location-estimation system, in contrast, the location estimation can be just an add-on feature to an existing system, and the wireless network can be used for other types of activities such as data communication and control as well. RF-based positioning systems are also found to be more suitable for large-scale deployments compared to infrared-based positioning systems.

The location-estimation systems developed using short-range wireless networking are sometimes referred to as *local positioning systems* (LPSs) to differentiate them from *global positioning systems* (GPSs). A GPS-enabled device determines its location by calculating its distance from three or more GPS satellites orbiting the Earth. Each GPS satellite continuously transmits a message containing the satellite location and the exact time. This message travels approximately with the speed of the light to reach the GPS receiver. The GPS receiver compares the exact time the message was received with the time the message was transmitted by the satellite to calculate the distance traveled. Knowing the distance to at least three satellites and the satellites' positions, the receiver calculates its own position. The LPS, in contrast, does not use information provided by GPS satellites or any other long-range transmitter. An LPS uses the RF signals transmitted by local nodes with known positions or the mobile node itself to calculate the location of the mobile node relative to the known locations of other local nodes.

The choice of location-estimation algorithm depends on the application scenario. The location-estimation methods are compared based on their performance and complexity. The location accuracy, which is the distance between the actual location and the estimated location, is the most intuitive performance metric. There is a major difference between implementing a locationing system with accuracy in centimeter range versus a system where a location-estimation error as high as several meters is still acceptable. If the goal of the location estimation is to track personnel in an office building, the coarse accuracy of several meters may be sufficient. The simple and low-cost locationing systems implemented based on short-range wireless networking protocols such as ZigBee are not targeted for very fine location estimation in centimeter range. Ultra-wideband (UWB) signals are more appropriate for very fine locationing.

The time it takes to determine the location of a node is another performance criterion. During the contention-based channel access period, for example, a ZigBee node may require several attempts before it gains access to the channel. If the objects are moving slowly, the latency to determine their location might not be a big concern. The coverage performance metric specifies the maximum physical space that an LPS can cover and successfully track the location of the nodes. Scalability is another metric that determines how well the locationing algorithm scales as the number of nodes increases and the coverage area is expanded. In low-cost applications, battery-powered nodes have limited computational power and small memory space. Therefore, these nodes might not be able to execute complex locationing algorithms.

The location estimation usually involves two groups of nodes. The first group consists of fixed nodes with known locations. These fixed nodes, sometimes referred to as *anchor nodes*, are used as references for the location estimation. The location of the anchor nodes can be determined by the installer, or the anchor nodes may be equipped with GPS to determine their own locations. If GPS is used as part of an LPS, the system is considered a hybrid GPS/LPS locationing system.

The second group is the nodes with unknown locations, referred to as *tracked nodes*. The main purpose of the location estimation is to determine the location of the tracked nodes with the help of the anchor nodes. The system should be able to keep on operating even if some anchor nodes are disabled. This is referred to as system *fault tolerance*. Some of the locationing algorithms can only determine the relative location of the tracked nodes to other tracked nodes instead of the anchor nodes.

The basic idea of local positioning can be summarized as follows. A tracked node with unknown location emits a signal, which is received by the neighboring anchor nodes. The anchor nodes measure the *received signal strength* (RSS), the *time of arrival* (ToA), or the *angle of arrival* (AoA) of the received signal. These measured values are used as inputs to an algorithm that determines the approximate location of the tracked node. The algorithms normally use only one of these three inputs. Measuring RSS is very simple and the ZigBee nodes are capable of measuring RSS for each received packet. Determining the precise time of arrival requires a very accurate clock. Finding the angle of arrival requires hardware modification and can increase the cost. The majority of the

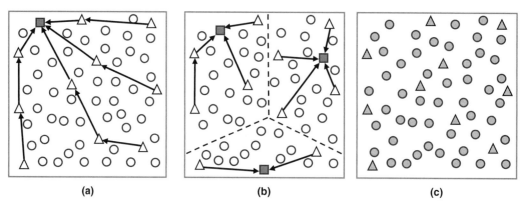

(a)   (b)   (c)

○ Tracked node that does not execute any location-estimation algorithm (but its location is tracked)

◉ Tracked node that executes (part of) the location-estimation algorithm

△ Anchor node that does not execute any location-estimation algorithm (but gathers and forwards RSSI)

▲ Anchor node that executes part of the location-estimation algorithm

■ The location processing node for the entire network or a section of the network

Figure 7.1: (a) Centralized, (b) Sectioned, and (c) Distributed Location Estimation Processing

RF-based locationing algorithms use only the RSS to estimate the location because of its simplicity and minimum or no hardware change requirement.

Figure 7.1 shows three scenarios for location estimation. In Figure 7.1a, a single node, referred to as the *central location processing node*, is dedicated to executing the location-estimation algorithm. All other nodes in the network only gather the location-related information such as RSS and send it to the central location processing node. This is referred to as a *centralized processing approach*. The central location processing node calculates the estimated location of all the tracked nodes and communicates the calculated location back to each tracked node if requested. The advantage of centralized processing is minimizing the required capabilities (e.g., processing power and memory space) of the nodes, except the central location processing node. The main disadvantages of this approach are creating high traffic levels and latency because all the nodes must communicate with a single node to determine their location. The high level of traffic can cause bottlenecks in the network and limit the location update rate. Also, if the next-hop routers of the central location processing node are battery powered, frequent relay of the locationing requests can drain their batteries rapidly. The latency and traffic problems get worse by increasing the size of the network. Therefore, the centralized processing method is more suitable for a small network or a network where the location update rate is low. The location update rate is the rate at which the estimated location of the nodes is updated (e.g., once every second).

One way to overcome the traffic bottleneck of the centralized processing method is to divide the network into sections and allocate a node capable of executing the locationing algorithm to each section. Figure 7.1b shows the nodes that are physically located in the same section communicating with their corresponding location processing nodes to determine their locations. An alternative approach, shown in Figure 7.1c, is to distribute the location-estimation task among almost all the nodes in the network. In this way, there is no centralized location processing node and each node determines its own location by communicating only with nearby anchor nodes and potentially other tracked nodes. In a fully distributed processing method, all the nodes must satisfy certain processing capabilities and memory space requirements. One of the advantages of distributed processing is relatively uniform packet traffic, which makes it easy to expand the size of the network.

Another figure of merit to compare different locationing methods is the required deployment density of the anchor nodes to achieve certain location accuracy. *Deployment density* is defined as the ratio of the number of anchor nodes to the coverage area. For example, if there are 20 anchor nodes covering $100\,\mathrm{m}^2$, the deployment density is $0.2\,\mathrm{nodes/m}^2$. For a given location-estimation accuracy, the lower the deployment density, the lower the implementation cost.

This chapter covers some of the locationing methods applicable to ZigBee wireless networking. However, the methods discussed in this chapter are only examples of the techniques that can be used for location estimation, and the locationing methods are not limited to these techniques.

7.2 Received Signal Strength-Based Locationing Algorithms

The received signal strength (energy) can be measured for each received packet. The measured signal energy is quantized to form the *received signal strength indicator* (RSSI). The RSSI and the time at which the packet was received (timestamp) are available to MAC, NWK, and APL layers for any type of analysis. For example, the simplest way to generate the link quality indicator (LQI) is to use the RSSI as an indication of link quality. The RSSI can also be used to develop a coarse but simple method of location estimation. Availability of RSSI means that a location-estimation system can be implemented without the need for any additional hardware for the individual nodes in the network.

There are four parameters associated with RSSI: dynamic range, accuracy, linearity, and averaging period. The RSSI dynamic range is specified in dB and indicates the minimum and maximum received signal energy that the receiver is capable of measuring. For example, if the RSSI provided by a receiver has dynamic range of 92 dB (from –88 dBm to +4 dBm), the minimum signal energy the receiver can measure is –88 dBm. Also, the maximum signal energy that this receiver can report as RSSI is 4 dBm. The RSSI accuracy indicates the average error associated with each received signal strength measurement. A typical commercially available transceiver is expected to be capable of providing ±4 dB or better typical RSSI accuracy. The RSSI linearity indicates the maximum deviation of the plot of RSSI from a straight line versus the actual received signal power (in logarithmic scale). The received signal strength is measured over a period of time and then averaged to generate RSSI. The averaging time is eight symbol periods, which is required by IEEE 802.15.4 if the RSSI is going to be used to generate LQI.

The simplest method to determine the location of a tracked node is to request that the tracked node transmit a signal. Then the location of the reference node that reports the highest RSSI is considered the estimated location of the tracked node. The advantage of this method is that it can be implemented easily on low-cost, battery-powered nodes with small memory size and low processing capabilities. However, the location-estimation accuracy of this method can be inadequate for many applications. The only way to improve the accuracy of this method is to increase the number of anchor nodes, which is not a desired approach in low-cost applications. The following section presents another simple RSSI-based locationing method.

7.2.1   RSSI-Based Location Estimation Using Trilateration

In an open environment with high probability of line-of-sight (LOS) and low-multipath effects, it might be possible to use a simple RSSI-based location-estimation algorithm if coarse accuracy is acceptable. Figure 7.2a shows an ideal location-estimation scenario where there are three nodes (nodes 1, 2, and 3) with known fixed locations. The fourth node is mobile, and the goal is to determine the estimated two-dimensional location of node 4. Two-dimensional (2D) means only X and Y coordinates of the node will be estimated. But the same concept can be extended to three-dimensional (3D) space as well. The location estimation in Figure 7.2a begins with node 4 transmitting a signal with a predefined output power. Assuming that all nodes in Figure 7.2a have omnidirectional

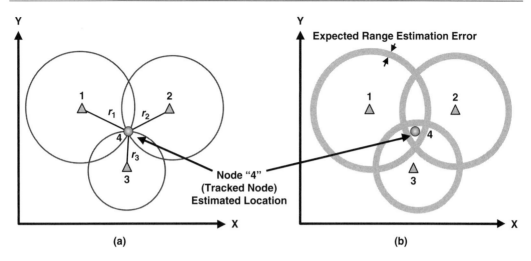

Figure 7.2: Location Estimation using Trilateration (a) Ideal Case and (b) with Range Estimation Error

antennas, each one of the fixed nodes 1–3 can estimate the distance (r) between its location and the location of node 4 using the following equation from Chapter 5:

$$P_R = P_T - 10 \times n \times \log_{10}(f) - 10 \times n \times \log_{10}(r) + 30 \times n - 32.44 \,(\text{dBm}) \qquad (7.1)$$

where P_T is the transmitted power (in dBm) by node 4, P_R is the RSS at the fixed node location, f is the transmitted signal frequency in MHz, n is the path-loss exponent, and r is the distance in meters.

Node 1, for example, can estimate the distance (r_1) between its location and the location of node 4 using RSS. From the single measurement done by node 1, the only conclusion that can be made is that node 4 is located on the perimeter of a circle with radius of r_1 centered at node 1. Using the Euclidian distance, we can write the following simple equation:

$$(X_1 - X_4)^2 + (Y_1 - Y_4)^2 = r_1^2$$

or:

$$(X_1 - X_4)^2 + (Y_1 - Y_4)^2 - r_1^2 = 0 \qquad (7.2)$$

where (X_1, Y_1) and (X_4, Y_4) are coordinates for node 1 and node 4, respectively. Similar equations can be derived for node 2 coordinates (X_2, Y_2) and node 3 coordinates (X_3, Y_3).

Therefore, to find the location of node 4, we need to find (X_4, Y_4) that satisfies the following equations:

$$\begin{bmatrix} (X_1 - X_4)^2 + (Y_1 - Y_4)^2 \\ (X_2 - X_4)^2 + (Y_2 - Y_4)^2 \\ (X_3 - X_4)^2 + (Y_3 - Y_4)^2 \end{bmatrix} - \begin{bmatrix} r_1^2 \\ r_2^2 \\ r_3^2 \end{bmatrix} = \begin{bmatrix} 0 \\ 0 \\ 0 \end{bmatrix} \qquad (7.3)$$

In the ideal scenario shown in Figure 7.2a, there will be a pair of coordinates (X_4, Y_4) that satisfies this equation. This method of determining the relative location of nodes using the geometry of triangles is referred to as *trilateration*. However, in a practical implementation, due to measurement errors it might not be possible to make the right-hand side of Equation 7.3 a true zero vector for any value of (X_4, Y_4). The RSSI provided by the transceivers has limited accuracies, which directly impacts the estimated distance between the nodes. The path-loss exponent is determined experimentally and can be a source of major error. As shown in Figure 7.2b, the circles associated with each fixed node might not even have a common intercept point when the actual range estimation error is higher than the expected range estimation error.

Since it is not feasible to make the right-hand side of Equation 7.3 a true zero, we can define an error vector (E) instead:

$$abs \left(\begin{bmatrix} (X_1 - X_4)^2 + (Y_1 - Y_4)^2 \\ (X_2 - X_4)^2 + (Y_2 - Y_4)^2 \\ (X_3 - X_4)^2 + (Y_3 - Y_4)^2 \end{bmatrix} - \begin{bmatrix} r_1^2 \\ r_2^2 \\ r_3^2 \end{bmatrix} \right) = \begin{bmatrix} e_1^2 \\ e_2^2 \\ e_3^2 \end{bmatrix} = E \qquad (7.4)$$

where *abs*(.) is the absolute value function.

If the square error is defined as:

$$\text{Square Error} = e_1^2 + e_2^2 + e_3^2 \qquad (7.5)$$

then the goal of location estimation becomes finding a pair of coordinates (X_4, Y_4) that minimizes the square error in Equation 7.4. This is a simple example of the classic optimization problem, where iterative or noniterative methods are used to minimize the value of an error function.

This simple RSSI-based location estimation can also be used when there are more than three fixed nodes with known locations. In this way the signal transmitted by the node with unknown location will be received by several nodes instead of only three nodes. The number of rows in Equation 7.4 is proportional to the number of fixed nodes participating in location estimation. Increasing the number of fixed nodes may improve the location-estimation accuracy in some applications. It is also possible to engage only the nearby nodes in location estimation. The RSSI value of the packet received by each anchor node indicates the distance between the nodes. If an anchor node receives a packet from the tracked node as part of the location-estimation process, the anchor node only participates in the location estimation if the RSSI of the received packet is above a certain limit. By modifying the RSSI limit, you increase or decrease the number of anchor nodes participating in the location estimation.

This simplified method was used in this subsection to describe the basic concept of location estimation using RSSI. This method requires further improvements to become a practical method of locationing. The error in RSSI measurement can result in an unacceptable level of inaccuracy in the estimated location. This method is not suitable for high multipath environments such as an office or an industrial warehouse. Alternative methods, which are more applicable to *indoor* location estimation, are discussed in Sections 7.2.3 and 7.2.4.

7.2.2   Sources of Error in RSSI-Based Location Estimation

The sources of error in RSSI-based location estimation can be divided into three main categories: hardware-related errors, the limitations of the location-estimation algorithm itself, and the effect of environment. This section briefly reviews the effect of some of these sources of error on location-estimation accuracy.

We can start with the uncertainty associated with the transmitted signal strength. The transmitter is expected to transmit a signal of a prespecified strength. However, the transceivers built for low-cost, short-range wireless networking normally have only a simple output power control mechanism, and the hardware manufacturers only guarantee a range of output power when a specific output power setting is selected. For example, a manufacturer may specify that the 0 dBm output power setting is associated with ± 2 dB of error. Not only might the amount of this error vary from one part to another, it also might depend on the ambient temperature as well. This specified error normally does not include the uncertainty due to the antenna.

The properties of the antenna itself can further degrade the accuracy of the transmitted signal power. As discussed in Chapter 5, the radiation pattern of an antenna is not omnidirectional and the transmitted signal strength at different directions can vary considerably. The antenna radiation pattern affects the accuracy of the signal strength measurements on the receiver side as well because the amount of attenuation due to the antenna will depend on the received signal angle of arrival. A simple location-estimation algorithm may not account for antenna radiation pattern variations. If the RSSI error is a constant known value and is not a function of environment or position of the node, it is possible to adjust the measured RSSI value accordingly by adding a known RSSI offset.

The receiver normally uses a simple mechanism to measure the received signal strength, and the difference between the actual received signal strength and the reported value is a source of error in location estimation. The received signal strength is quantized to form RSSI, and this quantization will be an unavoidable source of error for any RSSI-based location-estimation method.

The locationing algorithm itself may contribute to location-estimation error. For example, an optimization algorithm may converge to a local minimum instead of the global minimum. A *local minimum* is the lowest point of a function for a range of input values, but the local minimum may not be *the* lowest possible value of a function. Global minimum is *the* lowest value of a function for all possible inputs. Using a less-than-optimal result will induce an error in the location estimation that can be avoided by utilizing a better optimization method. The operations involving matrices may suffer from the presence of ill-conditioned matrices. An *ill-conditioned matrix* is a matrix for which the computations involving this matrix can be overly sensitive to the presence of small errors. In resource-limited applications, using complex algorithms to improve accuracy might not be an option.

The effect of the environment is normally the most challenging source of location-estimation error. The multipath can greatly affect the received signal strength. The mobility of the tracked node causes a dynamic change in the fading channel and makes it more challenging to compensate for the fading effect. Multipath mitigation methods such as antenna diversity or channel estimation methods can be used to help reduce errors due to channel fading. The signal spreading (e.g., direct sequence spread spectrum) can also reduce the multipath-induced error in some environments.

Increasing the period of RSSI measurement or repeating the RSSI measurement multiple times can reduce the error in some applications at the cost of reducing the network lifetime and degrading the location update rate. But, depending on the source of the error, increasing the number of measurements does not always improve the accuracy. For example, most hardware-related errors vary from one part to another and may depend on the ambient temperature as well.

There are two methods for estimating the channel characteristics: empirical or theoretical. In the *empirical approach*, an actual site survey is performed and the signal strength is measured at several locations and moments in time. These measured data points are used to create a model for the channel that includes the effect of fading as well. By repeating the RSSI measurement at a fixed location, it is possible to create a histogram of the RSSI values, which can be used to estimate the statistical model of the channel.

In the *theoretical approach*, the available information regarding the indoor environment that does not require any actual measurements is used to model the expected channel characteristics. Ray tracing is a popular method that uses the indoor environment floor plan, the reflection coefficients of the surrounding materials, and other information that is available without any site survey to compose a mathematical model for the channel. The advantage of the theoretical approach is that it eliminates the need for labor-intensive site surveys. The empirical method, if properly conducted, can have the advantage of improving accuracy beyond what the theoretical channel model can offer.

7.2.3   Location Estimation Based on Location Fingerprinting

Location estimation based on location fingerprinting is implemented in two phases. The first phase requires a site survey (offline training) to generate a database of measured RSSI values of the signals from the anchor nodes at certain locations. In the second phase (the real-time phase), each tracked node is capable of determining its own location by comparing the real-time measured RSSI of the signals received from the anchor nodes with the corresponding RSSI information available in its database [1].

The basic concept of this method is shown in Figure 7.3. The fixed nodes with known locations are numbered from 1 to 9. These fixed nodes (anchor nodes) have overlapping coverage and form a *grid*. The physical distances between the nodes are not necessarily equal. The smallest distance between two fixed nodes in a grid is known as the *grid spacing*. During the first phase, a receiver is placed at each predetermined location $L1$ to $L6$, and the RSSI of the received signals from fixed nodes 1 to 9 are measured and stored in an array. Typically, the array of the received signal strength in nonlogarithmic scale is represented by the *ss* notation. For example, at location $L1$, the array containing the received signal strength is the following:

$$ss_{L1} = [ss_{L1\,1}\ ss_{L1\,2}\ ...\ ss_{L1\,9}] \tag{7.6}$$

where $ss_{L1\,i}$ is the strength of the signal received from the anchor node i at location $L1$. The database containing the signal strength information associated with all locations $L1$

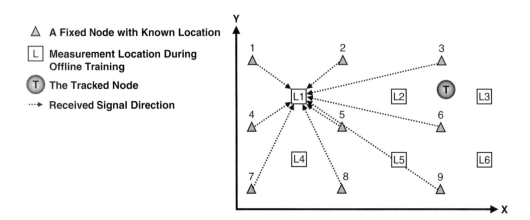

Figure 7.3: Location Estimation Based on Fingerprinting. The Fixed Nodes Act as Transmitters During the Offline Training

to *L6* is referred to as the *radio map*. In practice, the signal strength at known location *L1* to *L6* are measured multiple times, and the signal strength array contains the statistical average of the strength of the signals received from anchor nodes 1 to 9. The array of signal strength values at each location is known as the *fingerprint* (or *RF signature*) of that location.

The signal strength at a given location may vary significantly depending on the data collector orientation. In one orientation, the antenna may have line-of-sight (LoS) connectivity to an anchor node, whereas in the opposite orientation, the person's body can create an obstruction. It is possible to include the person's body orientation as part of the RF signature of the location or repeat the measurement with different orientation and average measured signal strengths. After the locationing system is implemented, the accuracy of the location estimation based on fingerprinting can be improved over time by gathering additional measured data points and adding them to the database created for the training phase.

After completion of the training phase, the tracked node in Figure 7.3 can determine its own location by going into receive mode and receiving the signals transmitted from each fixed node. The strength of each signal is calculated and stored in an array associated with the tracked-node current location:

$$ss_{current} = [ss_{current1} \ ss_{current2} \ \dots \ ss_{current9}] \qquad (7.7)$$

where $ss_{current\ i}$ is the strength of the signal received from the anchor node i at the current location of the tracked node. The Euclidian distance can be used to determine the distance (difference) between the current signal strength array measured during the real-time phase and the signal strength array associated with each known location. For example, the Euclidian distance between the ss_{L1} and $ss_{current}$ is calculated from the following equation:

$$d(ss_{current}, ss_{L1}) = \sqrt{(ss_{current1} - ss_{L1\ 1})^2 + (ss_{current2} - ss_{L1\ 2})^2 + ... + (ss_{current9} - ss_{L1\ 9})^2}$$

$$d(ss_{current}, ss_{L1}) = \sqrt{\sum_{i=1}^{9} (ss_{current\ i} - ss_{L1\ i})^2} \tag{7.8}$$

where $d(ss_{L1}, ss_{current})$ is the distance between these arrays. If the current location of the tracked node is at location $L1$, then, in theory, the arrays ss_{L1} and $ss_{current}$ are identical and their distance is zero ($d = 0$). However, in practice, the environment is dynamic and therefore these vectors will not be identical. In other words, at $L1$ location, the distance between ss_{L1} and $ss_{current}$ is small, but not zero. This distance is not a physical distance and is only an indication of the similarity of ss_{L1} and $ss_{current}$ signal strength arrays.

The simplest method for determining the location of the tracked node in the real-time phase is the *single nearest-neighbor* technique. In this method the tracked node calculates the Euclidian distance between the real-time measured signal strength array ($ss_{current}$) and the signal strength array (ss) associated with the locations $L1$ to $L6$. The location of the tracked node is simply estimated to be equal to one of $L1$ to $L6$ known locations, where the $ss_{L1\ i}$ array has minimum distance to the $ss_{current}$ array. The advantage of the single nearest neighbor method is its simplicity, but it does not take advantage of the available ss arrays associated with the rest of the known locations to improve the location-estimation accuracy.

The *k-nearest neighbor* (KNN) method, shown in Figure 7.4, can be used instead of the single nearest neighbor to improve the location-estimation accuracy. In this method, the tracked node identifies k known locations for which their ss array has the lowest distance to the $ss_{current}$ array at the current location of the tracked node. In the KNN technique, the estimated location of the tracked node is the average of these k known locations:

$$\begin{cases} X_E = \dfrac{1}{k}X_1 + \dfrac{1}{k}X_2 + ... + \dfrac{1}{k}X_k = \dfrac{1}{k}\sum_{i=1}^{k} X_i \\[2mm] Y_E = \dfrac{1}{k}Y_1 + \dfrac{1}{k}Y_2 + ... + \dfrac{1}{k}Y_k = \dfrac{1}{k}\sum_{i=1}^{k} Y_i \end{cases} \tag{7.9}$$

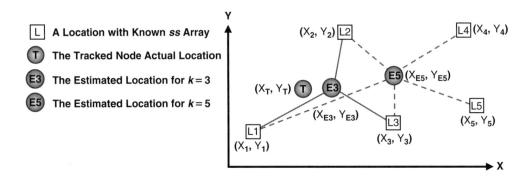

Figure 7.4: Effect of Increasing *k* in the *k*-nearest Neighbor (KNN) Method

where (X_E, Y_E) is the estimated location of the tracked node and (X_1, Y_1) to (X_k, Y_k) are the coordinates of the *k*-nearest neighbors. In other words, the tracked node estimated location is the location that minimizes the total Euclidian distance to these *k*-nearest neighbors.

For example, assuming that the location *L1* in Figure 7.4 has the *ss* array with the smallest distance to $ss_{current}$ array, and *L2* and *L3* have the next two closest arrays:

$$d(ss_{current}, ss_{L1}) < d(ss_{current}, ss_{L2}) < d(ss_{current}, ss_{L3}) \qquad (7.10)$$

then in the nearest-neighbor method, the *L1* will be the estimated location of the tracked node. In *k*-nearest neighbors with *k* = 3, the estimated location of the tracked node is *E3* in Figure 7.4, which is closer to the actual location of the tracked node compared to the estimate provided by the nearest-neighbor method.

Increasing the value of *k* will not necessarily improve the location-estimation accuracy. For example, in Figure 7.4 the estimated location when *k* is equal to 3 results in better estimation than *k* = 5. The reason is that by increasing the value of *k*, the further-away nodes are taken into account and may increase the estimation error.

The *weighted k nearest-neighbor* method can further improve the location-estimation accuracy of the KNN technique. In the KNN approach, all selected *k* neighbors (regardless of the distances of their associated *ss* arrays from the $ss_{current}$ array) are treated equally in determining the estimated location. Ignoring the differences between these neighbors can be a source of error because the tracked node may be closer to some neighbors than others and this information will be lost in simple averaging of the location of all *k* nearest neighbors to estimate the location of the tracked node. In the weighted

k nearest-neighbor method, the distance of ss array associated with each k nearest neighbor from the $ss_{current}$ is taken into account in estimating the location of the tracked node:

$$
\begin{cases}
X_E = \dfrac{1}{D}\left(\dfrac{1}{d(ss_{L1},ss_{current})+d_0}X_1 + \dfrac{1}{d(ss_{L2},ss_{current})+d_0}X_2 + \ldots \right. \\
\qquad\left. + \dfrac{1}{d(ss_{Lk},ss_{current})+d_0}X_k\right) \\
Y_E = \dfrac{1}{D}\left(\dfrac{1}{d(ss_{L1},ss_{current})+d_0}Y_1 + \dfrac{1}{d(ss_{L2},ss_{current})+d_0}Y_2 + \ldots \right. \\
\qquad\left. + \dfrac{1}{d(ss_{Lk},ss_{current})+d_0}Y_k\right)
\end{cases}
$$

$$
\begin{cases}
X_E = \dfrac{1}{D}\displaystyle\sum_{i=1}^{k}\dfrac{1}{d(ss_{Li},ss_{current})+d_0}X_i \\
Y_E = \dfrac{1}{D}\displaystyle\sum_{i=1}^{k}\dfrac{1}{d(ss_{Li},ss_{current})+d_0}Y_i
\end{cases}
\tag{7.11}
$$

$$
D = \sum_{i=1}^{k}\frac{1}{d(ss_{Li},ss_{current})+d_0}
\tag{7.12}
$$

In some instances, it is possible that the Euclidian distance between two vectors becomes zero. To avoid dividing by zero in Equation 7.12, a small (negligible) value of d_0 is added to all the denominators.

In the location estimation described previously, every time the location of the tracked node is being updated, the tracked node goes into receive mode and generates the signal strength (ss) array associated with its current location based on the signals it receives from the anchor nodes. Alternatively, the tracked node can act as a transmitter and all anchor nodes act as receivers. In this alternative method, shown in Figure 7.5, during the offline training a transmitter is placed at $L1$ location. The transmitted signal from this location will be received by the fixed nodes with known locations. Each fixed node measures the received signal strength, and the signal strength (ss) array associated with $L1$ location will be created (ss_{L1}). This is repeated for the rest of the selected locations ($L2$ to $L6$). In the second phase (real-time mode), every time there is a need to estimate the location of the tracked node, the tracked node transmits a signal that will be received by the fixed nodes 1 to 9. Then the signal strength (ss) array associated with the current

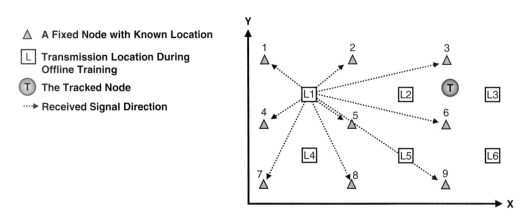

Figure 7.5: Alternative Approach in Location Estimation based on Fingerprinting. The Fixed Nodes Act as Receivers During the Offline Training

location of the tracked node is generated and used to estimate the location of the tracked node. Considering the fact that, in most environments, the RF propagation characteristics are symmetric, the accuracy of this alternative approach is expected to be the same as the previous scenario, where the tracked node acts as a receiver instead of a transmitter.

Statistical methods can be used to improve the accuracy of fingerprinting techniques. The *Bayesian filter* method, for example, is a way of probabilistically estimating a location from noisy observations. The *Kalman filter*, a variant of the Bayesian filter, is widely used in location estimation and tracking due to its computational efficiency.

The nearest-neighbor(s) mechanisms discussed here are not the only database search mechanisms. Several methods can be used to compare an array (or a vector) against a database of arrays. The basic concept of location estimation using fingerprinting can be seen as providing a database of known information to a system and expecting the system to *learn* how to relate the RSSI information to a specific physical location. Therefore, the algorithms developed for other disciplines such as machine learning, neural networks, and pattern recognition can be used for fingerprinting-based location estimation as well. But many of these algorithms are not necessarily developed for resource-limited nodes that are common in low-cost ZigBee networks. Some of these algorithms require a large database to learn properly, which is a burden in practical implementations. The available memory space, processing capabilities, and expected battery life of the wireless nodes should be used to determine the best algorithm for each application scenario.

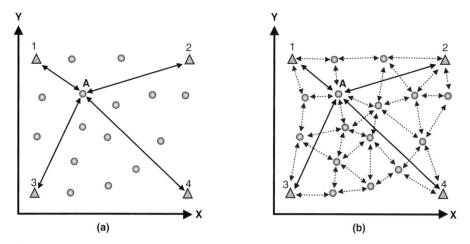

△ A Fixed Node with Known Location (Anchor Nodes)

◎ A Tracked Node

◄––► Estimated Distance Between a Tracked Node and an Anchor Node

◄····► Estimated Distance Between Two Tracked Nodes

Figure 7.6: (a) Basic Trilateration Method and (b) Cooperative Location Estimation

7.2.4 Cooperative Location Estimation

In cooperative location estimation, not only are the distances from the tracked node to the anchor nodes measured but also the relative distances of the tracked nodes to each other are used as part of location estimation. Figure 7.6 highlights the difference between the basic trilateration method and the cooperative technique. In Figure 7.6a, the location of the tracked node A is determined using range estimation between the node A and the anchor nodes 1 to 4. The other tracked nodes with unknown locations do not participate in determining the location of the tracked node A. Every time a node needs to determine its own location using trilateration, only the tracked node itself and the nearby anchor nodes will participate in locationing.

In the cooperative method shown in Figure 7.6b, in contrast, the location of several tracked nodes can be determined concurrently using an iterative method. First, the RSSI measurements at the anchor nodes provide an estimate for the location of the tracked nodes participating in cooperative locationing. Then each tracked node determines its approximate distance to the neighboring tracked nodes using the RSSI. The approximate distance between the tracked nodes is the additional information available in the cooperative method, which helps refine the location-estimation accuracy beyond the achievable accuracy in a basic trilateration method. The cooperative location-estimation involves nonlinear optimization and can become computationally intensive. In the cooperative method, the estimated locations of the tracked nodes are iteratively adjusted until the optimum result is achieved.

In a trilateration method, increasing the number of anchor nodes in a given area results in an improvement in location accuracy. But increasing the number of tracked nodes does not have any positive effect on accuracy of the trilateration technique. In the cooperative method, on the other hand, increasing either the number of anchor nodes or the number of tracked nodes can result in improvement in location-estimation accuracy.

7.3   Angle-of-Arrival-Based Algorithms

When a transceiver equipped with a simple antenna such as quarter-wave monopole antenna receives a signal, it can measure the signal strength, but the transceiver is incapable of determining the direction from which the signal has arrived. At the expense of additional complexity and cost, it is possible to modify a node to become capable of determining the received signal *angle of arrival* (AoA). For example, if the wireless node uses an antenna array, the signal AoA is available and can be used for location estimation instead of RSSI.

Assuming that tracked node is capable of determining received signal AoA, a simple algorithm can be used to estimate the location of the tracked node. Figure 7.7 describes the basic concept of location estimation based on AoA. The anchor nodes, with known locations, transmit signals using omnidirectional antennas. The tracked node receives the signals from the nearby anchor nodes and can estimate the received signal AoA. If the tracked node knows its own orientation, only two anchor nodes are required to determine the location of the tracked node. A node knows its orientation if it is aware of the direction North or a direction commonly known by the anchor nodes and the tracked node. Figure 7.7 shows a scenario in which the tracked node is unaware of its own orientation and therefore must receive the signal from at least three anchor nodes to be able to determine its own location. The uncertainty of the received signal AoA is shown by a gray shade in Figure 7.7. Although the tracked node does not know its orientation, it can calculate the angle between nodes 1, 4, and 2 in degrees:

$$1\hat{4}2 = 360° - (\theta_1 - \theta_2) \tag{7.13}$$

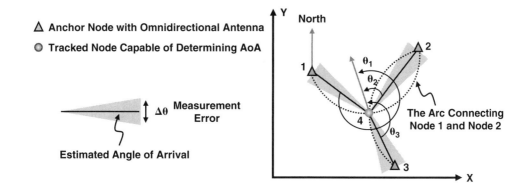

Figure 7.7: Location Estimation using Angle of Arrival (AoA)

If the tracked node knows the AoA for only nodes 1 and 2, the location of the tracked node can be anywhere on an arc connecting nodes 1 and 2. By measuring the AoA of the signal received from anchor node 3, the tracked node can calculate the angle between nodes 2, 4, and 3 as well:

$$2\hat{4}3 = (\theta_3 - \theta_2) \tag{7.14}$$

Since the locations of the anchor nodes are known, node 4 (the tracked node) can determine the arcs corresponding to its angle with nodes 1, 2 and, 3. The intercept of the two arcs, shown in Figure 7.7, is the location of node 4.

In AoA-based locationing, either the tracked nodes or the anchor nodes must be capable of determining the AoA. The AoA-based location estimation has some drawbacks compared to the RSSI-based locationing. In addition to the need for special hardware for detecting the AoA, there should not be any major obstacle between the tracked node and the anchor nodes in AoA locationing. Some AoA algorithms assume either there is an LoS between the tracked node and the anchor nodes or the AoA have certain distributions centered on the direction of LoS. These assumptions limit the use of AoA-based locationing in some indoor application scenarios.

7.4 Time-Based Algorithms (ToA and TDoA)

The time-based and RSSI-based locationing algorithms have a common goal: determining the distance between the nodes based on the properties of the received signal. In the

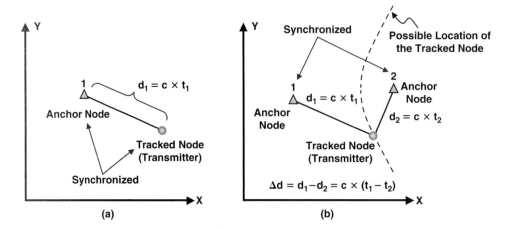

Figure 7.8: (a) Estimating the Distance using ToA and (b) Determining the Δd Based on TDoA

RSSI-based method, the received signal strength and the path-loss properties of the environment are used to estimate the distance. In time-based locationing algorithms, the estimated propagation speed of the signal and the time it takes for the signal to travel from the transmitter to the receiver are used to determine the distance between the nodes. GPS is an example of time-based locationing.

A time-based location estimation can be based on either the received signal time of arrival (ToA) or the time difference of arrival (TDoA). The ToA, shown in Figure 7.8a, requires synchronization between the receiver and transmitter. The ToA is the absolute value of the signal time of flight from the transmitter to the receiver. The distance from the tracked node to the anchor node (d_1) can be derived from the ToA (t_1) and the propagation speeds (e.g., $c =$ speed of light). The TDoA, in contrast, requires only synchronization of the receivers. The anchor nodes receive the signal transmitted by the tracked node, and the difference between the signal arrival times at these two anchor nodes can be used to calculate the Δd, which is the difference between the distance of d_1 and d_2. The TDoA requires participation of at least three anchor nodes to locate the position of the tracked node.

For the nodes shown in Figure 7.8b, we can write the following equations:

$$d_1 = \sqrt{(X_1 - X_E)^2 + (Y_1 - Y_E)^2} \tag{7.15}$$

$$d_2 = \sqrt{(X_2 - X_E)^2 + (Y_2 - Y_E)^2} \qquad (7.16)$$

$$\sqrt{(X_1 - X_E)^2 + (Y_1 - Y_E)^2} - \sqrt{(X_2 - X_E)^2 + (Y_2 - Y_E)^2} = \Delta d \qquad (7.17)$$

where (X_1, Y_1), (X_2, Y_2), and (X_E, Y_E) are the coordinates of anchor node 1, anchor node 2, and the estimated location of the tracked node, respectively. If only two anchor nodes participate in TDoA locationing, the only conclusion that can be made from Equation 7.17 is that the tracked node is located on a hyperbolic curve, shown as a dashed line in Figure 7.8b. When a third anchor node is added, the estimated location of the tracked node will be intersection of the corresponding hyperbolic curves.

Synchronization is critical in time-based location estimation. The faster the signal travels in the air, the more significant the timing error will become. For instance, clock synchronization is more critical when an RF signal is used instead of ultrasound. Also, in an indoor environment, the effect of multipath can increase the estimation error considerably. Signal spreading is a method used to mitigate the effect of multipath. A very high-bandwidth signal such as a UWB signal improves the location estimation accuracy significantly beyond the accuracy achievable using an IEEE 802.15.4 signal. The systems using UWB for time-based location estimation can achieve resolution as low as a few inches. Generally speaking, when signals are using spreading techniques, the higher the bandwidth, the better the location estimation accuracy.

For indoor environments, time-based location estimation using RF signals is less likely than RSSI-based locationing to be used in ZigBee wireless networks. The small bandwidth of IEEE 802.15.4 signals, compared to UWB signals, might not be sufficient to improve the TOA error when there is no direct LoS.

7.5   The Computational Complexity

In selecting an algorithm, one of the considerations is its computational complexity. *Computational complexity* is the study of the amount of resources needed to complete a task or perform a task in a certain way. This study also determines how well the algorithm scales if the number of inputs (e.g., the number of tracked nodes) is increased. Generally, the size of the input is represented by n. If in an algorithm the number of steps required to complete the task increases with the ratio of n^2 as the number of inputs increases, the algorithm is considered to have an n^2 *complexity*. In other words, in this algorithm, the number of steps required to deliver the final result quadruples every time the number of inputs is doubled. The *big O notation* is used to describe the computational complexity of an algorithm. In the case of an n^2 complexity algorithm, for example, the computational complexity is shown by $O(n^2)$, where n is the number of inputs.

The computational complexity is formally defined as follows: The function $f(n)$ has the computational complexity of $O(g(n))$ if and only if there exists a constant M (independent of n) such that:

$$|f(n)| \leq M|g(n)|$$

$$\text{for all } n \geq n_0$$

$$\text{for some fixed } n_0 > 0$$

One of the properties of the $f(n) = O(g(n))$ is the following:

$$O(\log(n)) < O(n) < O(n \log(n)) < O(n^2) < O(n^3) < O(n^4)$$

This means that, for example, the computational complexity of an $O(n^2)$ algorithm is less than an $O(n^3)$ algorithm but more than $O(n \log(n))$ algorithm.

References

[1] P. Bahl and V. N. Padmanabhan, "RADAR: An In-Building RF-Based User Location and Tracking System," *IEEE INFOCOM*, March 2000, pp. 775–784.
[2] R. Battiti, M. Brunato, and A. Villani, "Statistical Learning Theory for Location Fingerprinting in Wireless LANs," Technical report DIT-02-0086, Università di Trento, Oct. 2002.
[3] R. Burda and C. Wietfeld, "Multimedia Over 802.15.4 and ZigBee Networks for Ambient Environment Control," *Proceedings of 2007 IEEE Vehicular Technology Conference*, pp. 179–183.
[4] National Animal Identification System (NAIS), available at www.usda.gov/nais.
[5] W. Park et al., "The Implementation of an Indoor Location System to Control ZigBee Home Networks," *2006 International SICE-ICASE Joint Conference*, pp. 2158–2161.
[6] R. Rong and M. L. Sichitiu, "Angle of Arrival Localization for Wireless Sensor Networks," *Proceedings of IEEE Sensor and Ad Hoc Communications and Networks*, Reston, VA, 2006, pp. 374–382.
[7] M. A. Youssef, A. Agrawala, and A. Udaya Shankar, "WLAN Location Determination via Clustering and Probability Distributions," *IEEE Conference on Pervasive Computing and Communications,* 2003, pp. 143–150.

CHAPTER 8

ZigBee Coexistence

Operation in a license-free frequency band brings the challenge of sharing the same frequency band with several other wireless networks. In most cases, there is no collaboration between these independent wireless networks, and operation of one network may adversely affect the others. This chapter reviews some of the methods used in IEEE 802.15.4 wireless networking that increase the robustness of the network against interference. Basic properties of IEEE 802.11b/g, Bluetooth, cordless phones, and microwave ovens are reviewed as examples of potential sources of interference to an IEEE 802.15.4 network.

8.1 Introduction

ZigBee nodes operate in frequency bands that are commonly used by some other wireless standards as well. For example, wireless Internet access in many residential and commercial buildings is provided by IEEE 802.11b/g (wireless LAN)-based systems operating in the 2.4 GHz ISM band. Bluetooth and some cordless phones also use the 2.4 GHz ISM frequency band. It is likely for these wireless devices to operate concurrently in close proximity to each other. ZigBee, IEEE 802.11, Bluetooth, and other wireless networking standards developed for operation in a shared frequency band are expected to have some level of tolerance for the presence of other wireless systems. *Coexistence* is defined as the ability to operate in proximity to other wireless devices [1]. In other words, a system with coexistence capabilities can perform a task in a given shared environment where other systems are performing their tasks using different sets of rules. The *coexistence mechanism* is the method used for reducing the interference of one system on another.

In analyzing the coexistence performance of a ZigBee network, not only is the robustness of the ZigBee network in terms of operation of other nearby systems studied, but the effect of the operation of the ZigBee system on other networks is also analyzed. Basic properties of a ZigBee network such as low RF transmission power, low duty cycle, and the CSMA/CA channel access mechanism help reduce the effect of the presence of a ZigBee wireless network on other nearby systems.

One possible source of interference for a ZigBee node is other ZigBee nodes transmitting concurrently in the same or adjacent frequency channels. Every ZigBee node performs CCA before initiating a transmission during the contention period and will avoid transmission if the channel is occupied. Exposed and hidden node problems, discussed in Chapter 3, are known shortfalls of the CSMA/CA mechanism. Similar to IEEE 802.15.4 systems, IEEE 802.11b/g nodes perform CCA before each transmission and can notice the presence of signals from other IEEE 802.11b/g nodes. But ZigBee signals have lower bandwidth and energy compared to typical IEEE 802.11b/g signals, and the CCA performed by an IEEE 802.11b/g node may declare a frequency channel available while the channel is still occupied by a ZigBee signal. In a beacon-enabled ZigBee network, the coordinator can allocate GTS and manage the flow of the packets to reduce the chance of packet collision. But a ZigBee network might not have any knowledge of the operating mechanism and active time periods of the nearby network that is based on a standard other than ZigBee.

The interfering signal does not have to be in the same frequency channel as the desired signal to cause performance degradation. For example, in the 2.4 GHz ISM band, a ZigBee receiver may choose to filter the signals outside the 2400–2483.5 MHz band before they reach the first stage of the receiver. But the interferers within the 2.4 GHz ISM band will reach the receiver without any filtering. Therefore, if there is an interfering signal with significantly higher power than the desired signal in a nearby frequency channel, the interferer can saturate the first stage of the receiver and reduce the chance of successful recovery of the desired signal. If the first stage of the receiver is linear enough to pass the desired signal along with the interferer to the baseband while minimizing the distortion to the desired signal, the baseband will filter out the interferer and the desired signal can be recovered properly.

An interfering signal that is present in the same frequency band as a ZigBee signal is referred to as an *in-band blocking* signal. For example, in the 2.4 GHz ISM band, an in-band blocking signal is one with a frequency of 2400 MHz to 2483.5 MHz. An in-band blocking signal might not be in the same frequency channel as the desired signal. In the 2.4 GHz ISM band, an interfering signal located outside the 2400 MHz to 2483.5 MHz frequency band is known as an *out-of-band blocking* signal. The effect of out-of-band blocking signals can be reduced by using a *band-select* filter. In the 2.4 GHz ISM band, for instance, the band-select filter allows the signals within the 2400 MHz to 2483.5 MHz band to pass through the filter with minimal attenuation, but out-of-band signals are considerably attenuated. Intuitively, the band-select filter does not help with in-band blocking signals. In low-cost ZigBee nodes, using a band-select filter might not be even an option due to cost or node size limitations.

There are two approaches to improving the coexistence performance of ZigBee networks: collaborative and noncollaborative. In collaborative methods, certain operations of the ZigBee network and the other network (e.g., an IEEE 802.11b/g network) are managed together. An example is discussed in Section 8.3, where a ZigBee network and an IEEE 802.11b/g network are synchronized; every time one network is active, the other network stays inactive to avoid packet collisions. In a collaborative method, there must be a communication link between the ZigBee network and the other network to implement and manage the collaboration. The noncollaborative methods are the procedures any ZigBee network can follow to improve its coexistence performance without any knowledge regarding the operating mechanism of the nearby interfering wireless devices. The noncollaborative methods are based on detecting and estimating interferences and avoiding them whenever possible. The noncollaborative methods that can be used in ZigBee wireless networking are reviewed in Section 8.2.

IEEE has formed the IEEE 802.19 Technical Advisory Group (TAG) that develops and maintains policies for IEEE 802 standards regarding coexistence [2]. Prior to IEEE 802.19 TAG, the IEEE 802.15 coexistence Task Group 2 (TG2) published the IEEE 802.15.2-2003 [3], which mainly focuses on coexistence of IEEE 802.11 (WLAN) and IEEE 802.15.1 systems.

Model-based simulations or empirical methods can be used to study the mutual effect of concurrent operations of two wireless networks with different standards and predict the PER degradation. An example of the simulated performance of IEEE 802.15.4 is provided in the specification [1].

8.2 ZigBee Noncollaborative Coexistence Mechanisms

This section reviews a number of the noncollaborative methods applicable to ZigBee wireless networking. Some of these methods, such as Carrier Sense Multiple Access with Collision Avoidance (CSMA/CA), are basic properties of the IEEE 802.15.4 Physical layer and apply to any use-case scenario. Other noncollaborative methods such as dynamic RF output power control are add-on features, and their corresponding algorithms can be added to the system by the manufacturer or the end user.

8.2.1 CSMA/CA Channel Access

An IEEE 802.15.4 node uses CSMA/CA before initiating any transmission except during an allocated guaranteed time slot (GTS). Performing CCA will determine whether the frequency channel is currently used by another device. Depending on CCA implementation, it is also possible to determine whether the occupying signal is from another IEEE 802.15.4-compliant device or from a device operating based on a different standard. The CSMA/CA is useful when the nearby wireless nodes operating on a different standard do not dynamically change their operation frequencies. If a frequency channel is clear during the CCA but becomes occupied when the IEEE 802.15.4 node starts transmitting, the concurrent use of the same frequency channel can cause performance degradation in both networks.

8.2.2 Extremely Low Duty Cycle

One of the main reasons ZigBee nodes can have very long battery life is their very low duty cycle. For example, a sensor node that wakes up every minute, performs CCA, and transmits its measured data may have a duty cycle of around 0.01%. This extremely low duty cycle means that a ZigBee node may create very low interference in other nodes. Also, the ZigBee node may need only a few milliseconds of channel availability to perform CCA and transmit its packets. Therefore, if the nearby network has frequent inactive periods of several milliseconds, ZigBee nodes can utilize these intervals to establish communication links and exchange packets.

8.2.3 Signal Spreading

As discussed in Chapter 4, a spreading method such as DSSS allows the desired signal to have the advantage of processing gain over any interferer that resides in the same frequency band. Therefore, the signal spreading generally improves the robustness of a network against interferers.

On the other hand, the signal spreading may reduce the interference caused by a ZigBee network to other networks as well. The total signal energy does not change by spreading, but when the same signal energy is distributed over a larger bandwidth after spreading, the signal energy per Hertz is reduced. For any wireless system operating near a ZigBee network, the energy of the ZigBee interferer within the frequency band of interest is more important than the total energy of the interfering signal. Reducing the interference energy per Hertz increases the signal-to-interference ratio (SIR) and improves the chance of signal recovery in the victim system.

8.2.4 Dynamic RF Output Power Selection

One of the methods for improving coexistence is to adjust the RF output power of the transmitter based on the channel condition and the distance between the nodes. Typically, the RF output power is set to the lowest level that corresponds to an acceptable level of communication reliability. Reducing the transmitter output power decreases the interference with other nearby wireless devices, but the recipient of the signal becomes more susceptible to the interference. If several attempts to deliver a packet have failed, the transmitter RF output power can be increased to improve the SIR, as discussed in Chapter 4. Increasing the signal power can improve the chance of successful packet delivery at the potential cost of increasing interference with other wireless nodes. The effect of transmitter output power on battery life was discussed in Chapter 6.

8.2.5 Mesh Networking and Location-Aware Routing

If a certain router node in the network is constantly in the presence of strong interferers that cause frequent failure in packet delivery to the next hop, a mesh network may have the option of selecting an alternative path to carry the message to the final destination and avoid the router located close to a major source of interference. This is sometimes referred to as *path diversity*. The routing mechanism and route repair methods were discussed in Chapter 3.

In location-aware routing, the information regarding the areas of possible high interference, if known, can be taken into account in calculating link-cost functions. In this way, the packet traffic flow is directed away from the high-interference areas whenever possible. But the interferers still affect the transmissions initiated by, or intended for, the nodes within the high-interference areas.

The application of directional antennas as part of a routing mechanism was discussed in Chapter 6 as a way of increasing routing energy efficiency. Directional antennas can also improve the coexistence performance of a network in some application scenarios. For example, using directional antennas as part of location-aware routing can further reduce the interference caused by a node to other nearby wireless devices because the majority of the transmitted signal energy is directed toward the intended destination compared to omnidirectional signal propagation. On the other hand, if a node equipped with a directional antenna is in receive mode, the interferences arriving from directions in which the antenna gain is low will be attenuated more than the signals coming from the desired direction. In this way, the directional antenna can improve the robustness of a node to some of the interferences.

8.2.6   Adjacent and Alternate Channel Performance

IEEE 802.15.4 adjacent and alternate channel requirements were discussed in Chapter 4. Although the jamming performance is measured when the adjacent or alternate channel is occupied by an IEEE 802.15.4 signal, these performance matrices can be used as a general indication of the receiver interference resistance. A receiver with a high adjacent and alternate channel rejection ratio can also tolerate the presence of strong interferences, even if the interferences are not IEEE 802.15.4-compliant signals. The jamming resistance of a receiver does not provide any insight regarding the performance of the receiver when the interferer is in the same frequency channel as the desired signal.

8.2.7   Frequency Channel Selection

Changing the frequency channel when the energy of the interferer signal in the desired channel is unacceptable can be a simple way of addressing the interference problem. ZigBee Pro provides frequency agility capability that allows the entire network to change channels in the face of interference. If the ZigBee network does not include the ZigBee Pro feature set, still the network can dynamically change the frequency channel whenever needed.

If the frequencies of operations and bandwidths of the interfering signals in the nearby networks are known, the frequency channel of the ZigBee network can be selected accordingly to minimize the effect of interfering signals. An example is provided in Section 8.3, where, despite the presence of an IEEE 802.11b/g network, there are certain frequency bands that stay unoccupied and can be used by the ZigBee network. This is referred to as *channel alignment*.

8.2.8   Adaptive Packet Length Selection

Another example of a noncollaborative method that can be used in a ZigBee network is adaptive packet length selection based on channel condition. Reducing the size of the packet is normally considered a way of improving the PER in presence of interferers. Generally speaking, a smaller packet has a better chance of receiving the destination before an interferer appears in the same frequency channel. However, some experiments have shown that reducing the packet length does not always result in better PER performance [4].

8.3   Coexistence with IEEE 802.11b/g

IEEE 802.11 (WLAN) is widely used in homes, offices, and public places to provide high-speed Internet access. IEEE 802.11b/g networks operate in the 2.4 GHz ISM band and can be the source of interference with IEEE 802.15.4 networks. An IEEE 802.11b/g node transmitter output power is typically between 12 to 18 dBm [1], but IEEE 802.11b/g nodes may have output power as high as 30 dBm. This is significantly higher than the typical 0 dBm output power of a ZigBee wireless node.

IEEE 802.11b signals have 22 MHz bandwidth after spreading using DSSS. The maximum data rate of an IEEE 802.11b node is 11 Mbps. The IEEE 802.11 g standard supports a data rate of up to 54 Mbps using orthogonal frequency division multiplexing (OFDM), while the frequency bandwidth of the signal is still 22 MHz. IEEE 802.11 g has backward compatibility with IEEE 802.11b.

A *packet collision* is defined as the event where one or more IEEE 802.11b signals corrupt an IEEE 802.15.4 packet transmission such that the retransmission of the IEEE 802.15.4 packet is required. The ZigBee nodes have lower output power, duty cycle, and signal frequency bandwidth compared to IEEE 802.11b/g systems; therefore, a ZigBee system is expected to have very little effect on the operation of an IEEE 802.11b/g system.

IEEE 802.11b/g standards define 14 overlapping channels in the 2.4 GHz ISM band. The center frequencies of the channels (f_C) are 5 MHz apart and the bandwidth of each channel is 22 MHz:

$$f_C = 2412 + 5 \times (k - 1)(\text{MHz})$$

$$1 \leq k \leq 11 \text{ in North America (FCC)}$$

$$1 \leq k \leq 13 \text{ in Europe (ETSI)}$$

(8.1)

where k is the channel number.

As shown in Figure 8.1, there are only three nonoverlapping IEEE 802.11b/g channels. In North America, the nonoverlapping channels are channel numbers 1, 6, and 11. In Europe, the nonoverlapping channels are 1, 7, and 13. The frequency bands in between the nonoverlapping IEEE 802.11b/g channels are referred to as the *guard bands*. If the ZigBee network is in proximity to an IEEE 802.11b/g network that only uses the nonoverlapping channels, the ZigBee network can use channel alignment to improve coexistence performance. In Figure 8.1, the ZigBee channels 15, 20, 25, and 26 are located between the IEEE 802.11b/g channels and will suffer the least from the IEEE 802.11b/g interference in North America compared to other ZigBee frequency channels.

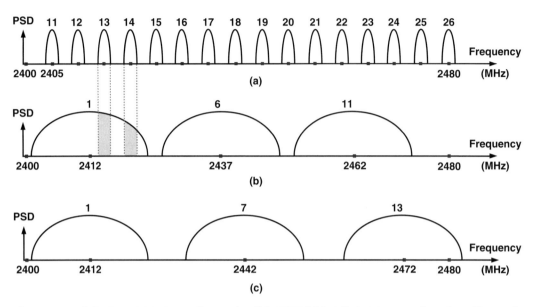

Figure 8.1: (a) IEEE 802.15.4 Channels, (b) IEEE802.11b/g Nonoverlapping Channels in North America, and (c) IEEE802.11b/g Nonoverlapping Channels in Europe

The ZigBee channels that are located in IEEE 802.11b guard bands might not enjoy an interference-free environment due to out-of-band emissions of 802.11b/g signals. Although the energy in these guard bands might not be zero, it will be lower than the signal energy within the IEEE 802.11b/g channels. Also, as discussed in Section 8.1, even if the interferer is not in the same frequency channel as the ZigBee signal, it can still degrade the PER performance of some ZigBee nodes.

In IEEE 802.11b, the signal energy is at its maximum at the center of the channel. Figure 8.1 shows that the ZigBee channel 13 has a lower SIR compared to channel 14. In IEEE OFDM-based 802.11 g, however, the power distribution over the channel is relatively flat. IEEE 802.11 g is widely used because of its superior throughput compared to IEEE 802.11b. For a ZigBee receiver, the DSSS modulated signal from the nearby WLAN appears as a wideband noise. A ZigBee signal is considered a narrowband interferer to a WLAN receiver and WLAN may use an adaptive notch filter to remove this narrowband interferer.

An IEEE 802.11b/g network may optionally use dynamic frequency selection to improve its coexistence performance if the network is subject to strong interferences. As a result of this dynamic frequency selection, the interference signals coming from IEEE 802.11b/g

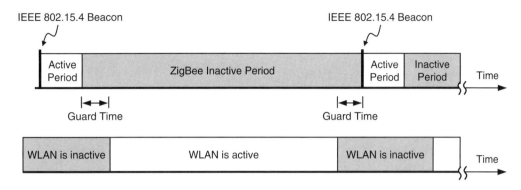

Figure 8.2: The Collaborative TDMA Method can be Used to Improve ZigBee and WLAN Coexistence

networks to a ZigBee network vary over time, and channel alignment might not be an option for the ZigBee network.

Both IEEE 802.11b/g and IEEE 802.15.4 use CSMA to gain access to a frequency channel. Both standards use CCA to determine if a channel is busy. An IEEE 802.15.4 node can detect the presence of an IEEE 802.11b/g signal, but the IEEE 802.11b/g node might not notice the presence of an IEEE 802.15.4 signal because of the small bandwidth and low signal power of typical ZigBee nodes.

The access point (AP) in an IEEE 802.11 system transmits beacon frames to synchronize the nodes in the networks. Since both IEEE 802.15.4 and IEEE 802.11 systems can operate in beacon-enabled mode, a time division multiple access (TDMA) method can be used to reduce the chance of interference between these two systems. The basic concept of this collaborative method is shown in Figure 8.2. The beacon-enabled operation of an IEEE 802.15.4 network is divided into active and inactive periods. The active period can be significantly shorter than the inactive period. Although the IEEE 802.15.4 system is active, the IEEE 802.11 stays inactive. The IEEE 802.11 resumes its activities after the IEEE 802.15.4 goes into inactive mode. There should be guard time between the active periods to account for possible clock inaccuracies.

This method requires a communication link between the ZigBee and WLAN systems to ensure that both systems are synchronized. One way to implement this collaboration is to have a dual-mode device in the network that contains transceivers for both IEEE 802.15.4

and IEEE 802.11b/g. This device will manage the time allocation to each network for TDMA operation.

8.4 Coexistence with Bluetooth

Bluetooth systems operate in the 2.4 GHz ISM band and use the frequency hopping spread spectrum (FHSS) method instead of DSSS to spread their signals. Figure 8.3 shows the Bluetooth basic operation mechanism. The transmitted signal bandwidth is 1 MHz, but the frequency channel is changed using a pseudorandom sequence. The maximum number of hops in Bluetooth is 1600 hops per second in the connection state. There are 79 frequency channels in Bluetooth separated by 1 MHz:

$$\text{Bluetooth frequency channels} = 2402 + k \quad 0 \leq k \leq 78 \tag{8.2}$$

where k is the channel number.

Bluetooth typical output power can be as high as 20 dBm. Bluetooth versions 1.1 and 1.2 were ratified as IEEE 802.15.1. But future versions of Bluetooth will not be ratified as any IEEE standard. The Bluetooth version 2.0 and higher can provide data rates of up to 3 Mbps. The device type in Bluetooth can be either master or slave. A master device can communicate with up to seven devices. The slaves periodically synchronize their clocks with the master.

FHSS, similarly to DSSS, provides processing gain, which improves the chance of successful packet delivery when interference is present. An IEEE 802.15.4 signal has 2 MHz bandwidth and may cause interference to three of the Bluetooth channels.

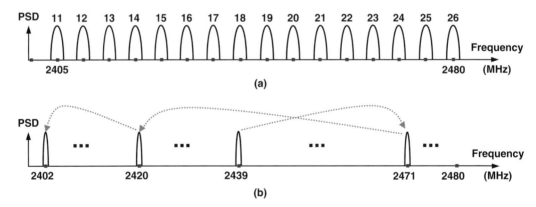

Figure 8.3: (a) IEEE 802.15.4 Channels and (b) Bluetooth Channels

Therefore, if the nearby Bluetooth device is using all 79 channels for frequency hopping, the maximum chance of interference between a single ZigBee node and a Bluetooth node is 3 out of 79 hops, which is approximately 4%.

However, a Bluetooth device can reduce the effect of presence of a ZigBee network (or any other network) by using adaptive frequency hopping (AFH). The AFH identifies the channels where interferences are present and marks these channels as "bad channels." Then the sequence of hops is modified such that the frequency channels with high-level interference are avoided. The bad channels in the frequency-hopping pattern are replaced with good channels via a lookup table. The Bluetooth master may periodically listen on a bad channel and if the interference has disappeared, the channel is marked as a good channel. Bluetooth slaves can also send a report regarding the channel quality to the master if necessary. The AFH method not only improves the performance of the Bluetooth network, it also reduces the effect of the Bluetooth network on other nearby networks that are not Bluetooth compliant.

The Bluetooth devices might not notice the presence of the ZigBee network due to the low duty cycle and low power of typical ZigBee nodes. If the frequency channel used by the ZigBee network is not marked as a bad channel, the Bluetooth network can cause interference to the ZigBee network, depending on the distance between the Bluetooth and ZigBee nodes.

8.5   Coexistence with Microwave Ovens

Microwave ovens use electromagnetic signals to generate heat in the material placed inside the oven (e.g., food or drink). The metal interior of the oven causes the signal to reflect from the walls of the oven and penetrate the food multiple times. Foods normally contain a high level of water, which has a very high attenuation constant and reduces the signal power considerably every time the signal passes through the food. The ovens may operate in the 2.4 GHz or sub-GHz frequency bands. The metal walls, ceiling, and floor of the oven prevent the majority of the RF signal energy from leaving the oven, but the electromagnetic leakage from the oven may still be sufficient to cause interference to other nearby wireless networks.

The center frequency of microwave ovens operating at 2.4 GHz is normally set at 2450 MHz. The signal bandwidth can be from 20 MHz to 80 MHz, depending on the microwave oven manufacturer. Typically, the IEEE 802.15.4 channels away from 2450 MHz have a better chance of avoiding performance degradation due to interference from microwave ovens.

A number of experiments have been performed to measure the signal strength outside several brands of microwave ovens [5]. These measurements are used to generate radiation pattern graphs for each individual oven. Oven radiation pattern depends on what is inside the oven (e.g., the type of material, size, and current orientation of the food) as well as oven manufacturer, oven nominal power, and orientation of the oven. The signal strength outside the oven is typically stronger in front of the oven compared to any other direction due to leakage from the oven door. The radiation outside an oven can be modeled by a transmitter with a nonuniform radiation pattern placed at the center of the oven. The peak of the transmitted signal strength of an oven can vary from –5 dBm to +20 dBm, averaged over 3 MHz bandwidth. The radiation from the oven is strong enough to be a source of interference to a nearby ZigBee wireless network.

The commercial microwave ovens may have duty cycles of close to 100% while operating. But the typical microwave ovens found in residential kitchens have duty cycles of around 50% or less. For example, when a microwave oven is turned on, the RF signal may only be present for 8 ms of every 20 ms. In some ZigBee applications, this 12 ms of oven inactivity can be sufficient for a ZigBee node that needs only a few milliseconds to perform CCA and transmit the packets to the destination.

8.6   Coexistence with Cordless Phones

Cordless phones operate in various frequencies, including the 915 MHz and 2.4 GHz frequency bands shared with ZigBee devices. Many of the cordless phones operate in 5.8 GHz and do not interfere with ZigBee networks. The cordless phones can implement DSSS or FHSS to improve their performance. Some DSSS cordless phones change their operation frequency channel only when the user presses a button on the phone itself. The exact characteristics of the cordless phone signals may vary from one manufacturer to another because normally the interoperability is not a concern for cordless phones.

The bandwidth of the signals from a DSSS-based cordless phone can be comparable to the bandwidth of a ZigBee signal. The transmitted signal strength from a 2.4 GHz cordless phone can be as high 30 dBm and becomes a major source of interference to a nearby ZigBee node if the frequency channels used by both devices are the same. Changing the frequency of operation of the ZigBee network can help reduce the effect of DSSS cordless phone interference. FHSS cordless phones use a spreading mechanism similar to Bluetooth and can cover the entire 2.4 GHz band. The transmitted signals from a cordless phone are normally stronger than a Bluetooth node. Packet collision between an FHSS cordless phone and a ZigBee network can cause some ZigBee nodes to retransmit their packets. A low-power and low-duty-cycle ZigBee network may have only a minor (or negligible) effect on cordless phone operation.

References

[1] "IEEE Std 802.15.4-2006: Wireless Medium Access Control (MAC) and Physical Layer (PHY) Specifications for Low-Rate Wireless Personal Area Networks (WPANs)," Sept. 2006.

[2] IEEE 802.19 Coexistence Technical Advisory Group (TAG), available at www.ieee802.org/19/.

[3] "IEEE 802.15.2 IEEE Recommended Practice for Information Technology Part 15.2: Coexistence of Wireless Personal Area Networks with Other Wireless Devices Operating in Unlicensed Frequency Bands," Aug. 2003.

[4] N. Golmie, "*Coexistence in Wireless Networks, Challenges and System-Level Solutions in the Unlicensed Bands*," University Press, Cambridge, 2006.

[5] P. E. Gawthrop, et al., "Radio Spectrum Measurements of Individual Microwave Ovens," NTIA Report 94-303-1, March 1994.

[6] J. Gutierrez, et al., "Low-Rate Wireless Personal Area Networks," *IEEE Press*, 2007.

[7] I. Howitt and J. A. Gutierrez, "IEEE 802.15.4 Low Rate: Wireless Personal Area Network Coexistence Issues," *IEEE Wireless Communications and Networking Conference*, March 2003, pp. 1481–1486.

CHAPTER 9

Related Technologies

ZigBee is not the only standard that adopts the IEEE 802.15.4 PHY and MAC layers to offer a wireless networking protocol. This chapter provides an overview of two other standards (6LoWPAN and WirelessHART) that reuse the IEEE 802.15.4 PHY and MAC layers as part of their wireless networking protocol. This chapter also reviews the basics of two wireless networking standards that are not based on IEEE 802.15.4 and compete with ZigBee in some application scenarios (Z-wave and ULP Bluetooth).

9.1   IPv6 over IEEE 802.15.4 (6LoWPAN)

The Internet Protocol (IP) version 6 (IPv6) [1] is a protocol developed by the Internet Engineering Task Force (IETF) [2]. IP is a network layer protocol for communication of data packets in a wired or wireless network. IP is the protocol used in the public Internet and many commercial networks. IPv6 replaces IP version 4 (IPv4). One of the major improvements in IPv6 compared to IPv4 is the address space. IPv4 supports 32-bit addressing, which translates to approximately 4.3 billion unique addresses. Although this might seem like a large address space, it will not be sufficient in the near future based on the expected growth rate of the nodes that will be connected to the Internet around the globe. IPv6 fixed this issue by supporting 128-bit addressing instead of 32-bit addressing in IPv4. This significant increase in the address space will ensure availability of unique addresses for all nodes in any practical growth-rate scenario.

In implementing a wireless sensor network, if the packet format is kept the same as the IP packet format, the interface between the wireless sensor network and the Internet can become simpler. This was the motivation behind development of the 6LoWPAN (IPv6 over low-power WPAN) standard. The 6LoWPAN is a standard that allows transmission of IPv6 packets over an IEEE 802.15.4 network.

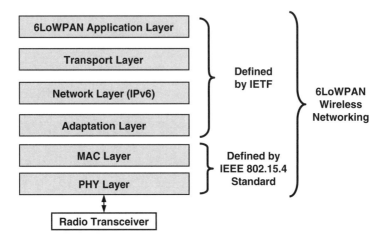

Figure 9.1: Protocol Layers in 6LoWPAN

Figure 9.1 shows the protocol layers in 6LoWPAN. The PHY and MAC layers are defined by IEEE 802.15.4. The packet format is different in IEEE 802.15.4 and IPv6 standards. The adaptation layer is created on top of the MAC layer to adapt IEEE 802.15.4 packets to IPv6, and vice versa. The next higher layer in 6LoWPAN is the IPv6 Network layer, which is compatible with any other IPv6 network regardless of its physical layer. The protocol layers above MAC are defined by IETF. The protocol layers in Figure 9.1 indicate that the user at the application layer always receives and transmits IPv6-compliant packets, whereas the packets transmitted over air are in IEEE 802.15.4 format. The 6LoWPAN standard, similar to ZigBee, supports mesh networking.

The 6LoWPAN standard is developed specifically for the nodes that have limited memory space and processing capabilities. The IPv6 requires support of packet sizes much larger than the largest IEEE 802.15.4 packet size. The size of the header in IPv6 is 40 octets. The 6LoWPAN uses a header compression method to reduce the size of the IPv6 packet header. But this compression is not sufficient to bring the size of IPv6 packets anywhere close to IEEE 802.15.4 packets. The maximum physical layer packet size in IEEE 802.15.4 is 127 octets (*aMaxPHYPacketSize*). The maximum size of the payload is less than 127 octets because of all necessary overheads in a packet. The *minimum* IPv6 packet size is 1280 octets, which is much larger than the *maximum* packet size in IEEE 802.15.4. Therefore, the IPv6 packets need to be fragmented by the transmitter and reassembled by the receiver to accommodate the requirements of both standards. The adaptation layer in 6LoWPAN is responsible for fragmenting and reassembling the packets.

One of the expected advantages of 6LoWPAN is the interoperability of the network with all other IP network links (wired and wireless). A 6LoWPAN node is capable of communicating with other IP-enabled devices. A wireless node that implements 6LoWPAN can be accessed and managed similarly to any other IP device. A user who is familiar with IPv6 can use tools and resources developed for IPv6 to implement its wireless sensor networking application while minimizing the interactions with lower layers of the protocol (IEEE 802.15.4).

6LoWPAN supports both 64-bit extended addressing and 16-bit short addressing in IEEE 802.15.4, but 6LoWPAN imposes additional constraints (beyond IEEE 802.15.4) on the format of the 16-bit short addressing. For example, in multicasting, the first three bits of the short address must be 100. This leaves 13 bits for the actual multicast address in a 6LoWPAN network. ZigBee, in contrast, uses all 16 bits for addressing.

6LoWPAN provides an alternative way of implementing a wireless network. The battery life in the 6LoWPAN and ZigBee standards should be comparable because of the similarities of their hardware and bottom two layers of their protocols. This is based on the assumption that both standards have comparable routing efficiency and are tested in similar use-case scenarios. The decision to select one standard versus another is determined by the target application. Consider an application for which there is no need to interface with IP-enabled devices and the average size of the packets is small. In this case, it is not necessary to implement 6LoWPAN, which performs fragmentation and reassembly of the packets to ensure their compatibility with IPv6.

9.2   WirelessHART

Highway Addressable Remote Transducer (HART) is a communications protocol for applications such as process control, equipment and process monitoring, advanced diagnostics, and closed loop control in wired industrial networks. HART supports a data rate of 1.2 Kbps using FSK modulation. HART uses a master/slave mechanism, and a slave device only transmits data when it is asked by a master device. HART is widely used in process control applications, but it is limited to wired networks.

WirelessHART is a wireless networking standard based on HART that adds wireless flexibility to an existing HART network. WirelessHART operates at the 2.4 GHz ISM band and is backward compatible with existing HART devices, commands, and tools [3]. WirelessHART supports mesh networking. For security, WirelessHART uses AES-128 block ciphers similar to the ZigBee standard.

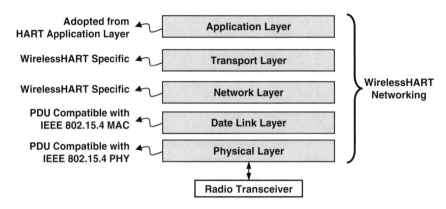

Figure 9.2: Protocol Layers in WirelessHART

The WirelessHART standard defines the protocol layers shown in Figure 9.2. The PHY layer of WirelessHART uses packet data units (PDUs) that are compatible with IEEE 802.15.4 PHY at 2.4 GHz. WirelessHART uses the same frequency channels as IEEE 802.15.4 with O-QPSK/DSSS modulation and supports a data rate of 250 Kbps. The difference between WirelessHART PHY and IEEE 802.15.4 PHY is that WirelessHART PHY hops over 16 channels defined by IEEE 802.15.4 on a packet-by-packet basis. WirelessHART can avoid hopping into certain channels by placing these channels in a "blacklist." WirelessHART also performs CCA before each transmission to avoid creating or experiencing interferences. WirelessHART can be implemented on the commercially available radios developed for the IEEE 802.15.4 standard because WirelessHART and IEEE 802.15.4 PHYs are compatible. The nominal transmitted power in WirelessHART is 10 dBm compared to 0 dBm (typical) in a ZigBee network.

The PDU in the Data Link layer of WirelessHART is compatible with IEEE 802.15.4 MAC PDU. WirelessHART uses a superframe to provide TDMA for communications between network devices. There are 100 time slots per second in WirelessHART. The nodes in WirelessHART may have dedicated time slots or use a contention-based channel access mechanism. The network layer is capable of mesh networking and supports broadcast, multicast, and unicast transmissions. The devices in the network maintain a record that includes information such as received signal strength from neighbors and the list of discovered neighbor devices.

The transport layer supports both acknowledged and unacknowledged communications and performs automatic retries a limited number of times if the initial data transmission is not successful. The WirelessHART application layer is based on the HART application layer to ensure compatibility of WirelessHART nodes with a HART network.

Both ZigBee and WirelessHART can be used in industrial monitoring and control applications. If a wired HART network is present, WirelessHART can be a better choice than ZigBee because of backward compatibility with the HART network. For typical wireless sensor networking applications where no interface with the HART network is required, ZigBee can be used instead of WirelessHART.

9.3   Z-wave

Z-wave is a wireless networking protocol developed by Z-wave alliance for 900MHz ISM band operation [4]. Unlike ZigBee, Z-wave defines all protocol layers and does not adopt IEEE 802.15.4 PHY and MAC layers. Z-wave supports 9.6Kbps and 40Kbps data rates using frequency shift keying (FSK) modulation. The Z-wave signals are narrowband, and no spreading method such as DSSS or FHSS is used to help Z-wave mitigate the presence of interferences and multipath nulls.

One of the differentiating factors between ZigBee and Z-wave is address space. ZigBee supports 64-bit and 16-bit addressing, whereas Z-wave supports only 8-bit addressing. Therefore, in a single Z-wave network, there can be up to 232 nodes. This can be sufficient in many applications. However, if a larger number of nodes is required, a ZigBee network can be a better alternative because even in 16-bit short addressing, there can be up to 65,536 nodes in a single ZigBee network. In Z-wave, each network is identified by a 32-bit value called *HomeID*.

Z-wave, similar to ZigBee, supports mesh networking, broadcasting, and multicasting. The collision avoidance in a Z-wave network is achieved by making sure the channel is available before each transmission. If the channel is not available, the node will use a random back-off mechanism between transmission attempts.

For security, Z-wave relays on the Triple Data Encryption Standard (TDES) [5]. The Data Encryption Standard (DES) is considered to be insecure for many applications due to small key size (56 bits). Triple DES uses DES three times to improve security. DES is superseded by the Advanced Encryption Standard (AES). The size of the key in AES can be 128 bits. ZigBee uses AES-128 to ensure communication security in the network.

In summary, a ZigBee network operating in the same sub-GHz frequency band as Z-wave can support higher (or comparable) data rates, a larger number of nodes, and a superior security method than a Z-wave network. Z-wave nodes have lower complexity and have the potential to cost less than a comparable ZigBee node.

9.4   Ultra-Low-Power Bluetooth (Wibree)

The Ultra-Low-Power (ULP) Bluetooth standard [6], originally known as Wibree [7], is a short-range wireless networking standard developed for point-to-point and very low-duty-cycle wireless communications. The ULP is a simplified version of the Bluetooth standard that expects to have an order of magnitude longer battery life compared to a typical Bluetooth device. For example, ULP has only one packet type, whereas Bluetooth has 28 packet types. ULP does not support mesh networking and therefore does not compete with ZigBee in applications that require mesh networking. ULP and ZigBee compete in short-range point-to-point wireless networking applications. ULP operates in the 2.4 GHz ISM band and defines 40 channels with 2 MHz channel spacing. The signal bandwidth is 1 MHz.

The Bluetooth standard is a well-established wireless networking protocol. ULP-enabled devices can communicate with a Bluetooth-enabled device only if the Bluetooth-enabled device has implemented a dual-mode Bluetooth/ULP protocol stack (see Figure 9.3).

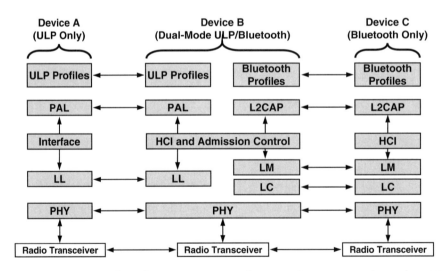

Figure 9.3: Interactions between Protocol Layers in ULP, Dual-Mode ULP/Bluetooth, and Bluetooth Devices

This dual-mode protocol stack allows the dual-mode device to communicate with both traditional Bluetooth devices and ULP nodes. The common features between ULP and traditional Bluetooth standards simplify the implementation of the dual-mode protocol stack. The cost (e.g., additional memory space) of upgrading a Bluetooth device to a dual-mode ULP/Bluetooth device should be small.

In Figure 9.3, device A contains only a single-mode ULP stack. The PHY layer in ULP is the same as traditional Bluetooth. The next higher layer above PHY is the Link Layer (LL). In Bluetooth there is a separate link controller (LC) and a link manager (LM). The ULP replaces both the LC and LM with a simple link layer (LL). The next upper layer in ULP is PAL, which is a subset of the Logical Link Control and Adaptation Protocol (L2CAP) available in Bluetooth. The quality of service (QoS), segmentation, and reassembly of packets are examples of duties performed by the L2CAP. The ULP, similar to ZigBee, does not support QoS. The protocol layers in device A can communicate with their corresponding layers in a dual-mode device (device B). Device C has only a Bluetooth stack and cannot communicate directly with device A. In the Bluetooth stack, there is a host controller interface (HCI) below L2CAP. The ULP uses a simpler interface mechanism.

9.5   TinyOS

An operating system (OS) in a large network provides an interface for a user (e.g., a programmer) to manage the resources in a network. Typical operating systems that are developed for low-power wireless sensor networks may require large memory space and high-performance microprocessors, which can be beyond the capabilities of resource-limited wireless sensor nodes. TinyOS is an open-source operating system from the TinyOS alliance that's specifically developed for low-power and resource-limited sensor networks in which the nodes spend the majority of their time in sleep mode [8].

TinyOS can be implemented as the operating system in a wireless sensor network where the PHY and MAC layers are defined by the IEEE 802.15.4 standard and the NWK and APL are implemented using the ZigBee standard. TinyOS is not limited to ZigBee networking and can be implemented as an operating system in other wireless sensor networking protocols as well. TinyOS provides a common programming environment for users. In ZigBee applications the users may choose to develop their own application-specific codes without using TinyOS. TinyOS uses the network embedded systems C (nesC) to develop the applications. nesC is a programming language developed specifically for battery-powered nodes with low memory capacity and processing capabilities.

References

[1] Ipv6 over IEEE 802.15.4 (6LoWPAN), available at http://6lowpan.net/.

[2] Internet Engineering Task Force (IETF), available at www.ietf.org.

[3] WirelessHART, available at www.hartcomm.org.

[4] Z-wave Alliance, available at www.z-wavealliance.org.

[5] Data Encryption Standard (DES), Federal Information Processing Standards Publication (FIPS PUB) 46-3, Oct. 1999, available at http://csrc.nist.gov.

[6] Bluetooth Standard, available at www.Bluetooth.com.

[7] Wibree, available at www.wibree.com.

[8] TinyOS Alliance, available at www.tinyOS.net.

PSSS Code Tables

A.1 PSSS Code Tables

The details of parallel sequence spread spectrum (PSSS)-based operation are provided in Chapter 4. In the PSSS approach, nearly orthogonal sequences are used to spread the signal before transmission. The PSSS sequences are identified by sequence numbers. Table A.1 presents the sequences used for operation in the 868MHz frequency band. Table A.2 contains the sequences for the 915MHz mode of operation. The length of each sequence is 32 bits.

Table A.1: PSSS Sequences

Sequence Number	Chip Number															
	0	1	2	3	4	5	6	7	8	9	10	11	12	13	14	15
0	-1	-1	-1	-1	1	-1	-1	1	-1	1	1	-1	-1	1	1	1
1	-1	1	-1	-1	-1	-1	1	-1	-1	1	-1	1	1	-1	-1	1
2	1	-1	1	-1	-1	-1	-1	1	-1	-1	1	-1	1	1	-1	-1
3	1	-1	1	-1	1	-1	-1	-1	-1	1	-1	-1	1	-1	1	1
4	1	1	-1	1	-1	1	-1	-1	-1	-1	1	-1	-1	1	-1	1
5	-1	1	1	1	-1	1	-1	1	-1	-1	-1	-1	1	-1	-1	1
6	1	-1	1	1	1	-1	1	-1	1	-1	-1	-1	-1	1	-1	-1
7	-1	1	1	-1	1	1	1	-1	1	-1	1	-1	-1	-1	-1	1
8	-1	-1	1	1	1	-1	1	1	1	-1	1	-1	1	-1	-1	-1
9	1	-1	-1	-1	1	1	-1	1	1	1	-1	1	-1	1	-1	-1
10	1	1	-1	-1	-1	1	1	-1	1	1	1	-1	1	-1	1	-1
11	1	1	1	1	-1	-1	-1	1	1	-1	1	1	1	-1	1	-1
12	1	1	1	1	1	-1	-1	-1	1	1	-1	1	1	1	-1	1
13	-1	-1	1	1	1	1	1	-1	-1	-1	1	1	-1	1	1	1
14	1	-1	-1	1	1	1	1	1	-1	-1	-1	1	1	-1	1	1
15	-1	1	1	-1	-1	1	1	1	1	1	-1	-1	-1	1	1	-1
16	1	-1	1	1	-1	-1	1	1	1	1	1	-1	-1	-1	1	1
17	-1	-1	1	-1	1	1	-1	-1	1	1	1	1	1	-1	-1	-1
18	1	-1	-1	1	-1	1	1	-1	-1	1	1	1	1	1	-1	-1
19	-1	-1	1	-1	-1	1	-1	1	1	-1	-1	1	1	1	1	1

Table A.2: PSSS Sequences

Sequence Number	Chip													
	0	1	2	3	4	5	6	7	8	9	10	11	12	13
0	-1	-1	-1	-1	1	-1	-1	1	-1	1	1	-1	-1	1
1	1	1	-1	1	-1	1	-1	-1	-1	-1	1	-1	-1	1
2	-1	-1	1	1	-1	1	1	1	-1	1	-1	1	-1	-1
3	1	1	1	1	1	-1	-1	-1	1	1	-1	1	1	1
4	1	-1	1	1	-1	-1	1	1	1	1	1	-1	-1	-1

for 868 MHz Operation

Chip Number															
16	17	18	19	20	21	22	23	24	25	26	27	28	29	30	31
1	1	-1	-1	-1	1	1	-1	1	1	1	-1	1	-1	1	-1
1	1	1	1	-1	-1	-1	1	1	-1	1	1	1	-1	1	-1
1	1	1	1	1	-1	-1	-1	1	1	-1	1	1	1	-1	1
-1	-1	1	1	1	1	1	1	-1	-1	-1	1	1	-1	1	1
1	-1	-1	1	1	1	1	1	-1	-1	-1	1	1	-1	1	1
-1	1	1	-1	-1	1	1	1	1	1	-1	-1	-1	1	1	-1
1	-1	1	1	-1	-1	1	1	1	1	1	-1	-1	-1	1	1
-1	-1	1	-1	1	1	-1	-1	1	1	1	1	1	-1	-1	-1
1	-1	-1	1	-1	1	1	-1	-1	1	1	1	1	1	-1	-1
-1	-1	1	-1	-1	1	-1	1	1	-1	-1	1	1	1	1	-1
-1	-1	-1	1	-1	-1	1	-1	1	1	-1	-1	1	1	1	1
1	-1	-1	-1	-1	1	-1	-1	1	-1	1	1	-1	-1	1	1
-1	1	-1	-1	-1	-1	1	-1	-1	1	-1	1	1	-1	-1	1
-1	1	-1	1	-1	-1	-1	-1	1	-1	-1	1	-1	1	1	-1
1	-1	1	-1	1	-1	-1	-1	-1	1	-1	-1	1	-1	1	1
1	1	1	-1	1	-1	1	-1	-1	-1	-1	1	-1	-1	1	-1
-1	1	1	1	-1	1	-1	1	-1	-1	-1	-1	1	-1	-1	1
1	1	-1	1	1	1	-1	1	-1	1	-1	-1	-1	-1	1	-1
-1	1	1	-1	1	1	1	-1	1	-1	1	-1	-1	-1	-1	1
-1	-1	-1	1	1	-1	1	1	1	-1	1	-1	1	-1	-1	-1

for 915 MHz Operation

Number																	
14	15	16	17	18	19	20	21	22	23	24	25	26	27	28	29	30	31
1	1	1	1	-1	-1	-1	1	1	-1	1	1	1	-1	1	-1	1	-1
-1	1	1	-1	-1	1	1	1	1	1	-1	-1	-1	1	1	-1	1	1
-1	-1	1	-1	-1	1	-1	1	1	-1	-1	1	1	1	1	1	-1	-1
-1	1	-1	1	-1	-1	-1	-1	1	-1	-1	1	-1	1	1	-1	-1	1
1	1	-1	1	1	1	-1	1	-1	1	-1	-1	-1	-1	1	-1	-1	1

ZigBee Device Profile Services

The device profile provides client and server services. Tables B.1, B.2, and B.3 summarize the commands supported by the client services. The mandatory and optional commands are identified by *M* and *O,* respectively. All the commands in these tables are optional for a client. Tables B.4, B.5, and B.6 are the commands supported by server services.

Table B.1: Device and Service Discovery (Client Services) Commands

Command (*ClusterID*)	Server	Description
NWK_addr_req (0x0000)	M	The local device (client) supplies the IEEE address of a remote device (server) and requests the NWK address associated with this IEEE address. This command can optionally request the NWK addresses of all the devices associated with the remote device.
IEEE_addr_req (0x0001)	M	The local device supplies the NWK address of a remote device and requests the IEEE address associated with this NWK address. This command can optionally request the NWK addresses of all the devices associated with the remote device.
Node_Desc_req (0x0002)	M	The local device requests the node descriptor of a remote device. The NWK address of the remote device is provided.
Power_Desc_req (0x0003)	M	The local device requests the node power descriptor of a remote device. The NWK address of the remote device is provided.

(Continued)

Table B.1: (Continued)

Command (*ClusterID*)	Server	Description
Simple_Desc_req (0x0004)	M	The local device requests the simple descriptor of a remote device. The NWK address and the endpoint address of the remote device are provided.
Active_EP_req (0x0005)	M	The local device requests a list of active endpoints at a given NWK address. An endpoint is active if it has a simple descriptor.
Match_Desc_req (0x0006)	M	The local device uses this command to find possible matched devices in another node. The local device supplies the NWK address of a remote device, the profile identifier, and the list of its own input and output clusters. The remote device will compare the local device clusters with its own clusters and reply with the list of endpoint addresses that have matched the request criteria.
Complex_Desc_req (0x0010)	O	The local device requests the complex descriptor of a remote device. The NWK address of the remote device is provided.
User_Desc_req (0x0011)	O	The local device requests the user descriptor of a remote device. The NWK address of the remote device is provided.
Discovery_Cache_req (0x0012)	M	The local device uses this command to locate a primary discovery cache device on the network. If the remote device that receives this request supports the *Discovery_Cache_req*, then the remote device will unicast a *Discovery_Cache_rsp* with the status of SUCCESS back to the local device.
End_Device_annce (0x0013)	O	This command contains the 16-bit NWK address, the 64-bit IEEE address, and the list of the capabilities of the end device that recently joined or rejoined the network. This command is broadcast to allow other devices in the network to know the NWK address of the new device as well as its capabilities.
User_Desc_set (0x0014)	O	The local device uses this command to configure the user descriptor on a remote device. The local device provides the NWK address of the remote device as well as the user descriptor.

Table B.1: (Continued)

Command (*ClusterID*)	Server	Description
System_Server_ Discovery_req (0x0015)	M	The local device broadcasts this command along with the 16-bit *ServerMask* to discover server(s) in the network.
Discovery_Store_req (0x0016)	O	The local device uses this command to request storage of its discovery cache information on a primary discovery cache device. If the remote device is a primary discovery cache device and has sufficient storage space, it will reply with the status of SUCCESS.
Node_Desc_Store_req (0x0017)	O	This command requests to store the node descriptor of the local device (end device) on a primary discovery cache device.
Power_Desc_Store_req (0x0018)	O	This command requests to store the power descriptor of the local device (end device) on a primary discovery cache device.
Active_EP_Store_req (0x0019)	O	This command requests to store the list of active endpoints of the local device (end device) on a primary discovery cache device.
Simple_Desc_Store_req (0x001a)	O	This command requests to store the list of Simple Descriptors of the local device (end device) on a primary discovery cache device.
Remove_node_cache_req (0x001b)	O	This command requests removal of any cache information stored on the primary discovery cache device related to a specified ZigBee end device.
Find_node_cache_req (0x001c)	M	This command is broadcast to find a device on the network that holds discovery information for a specific device. This command provides the NWK and IEEE addresses of the devices of interest.

Table B.2: Binding-related (Client Services) Commands

Command (*ClusterID*)	Server	Description
End_Device_Bind_req (0x0020)	O	The local device uses this command to perform end device bind with a remote device. This command is unicast to the ZigBee coordinator.
Bind_req (0x0021)	O	This command is used to create an entry in a binding table for two devices. The IEEE addresses and endpoint addresses of these two devices (source and destination) as well as the *clusterID* of the source device are provided as part of the command. This command is unicast to a primary binding table cache or the source device itself.
Unbind_req (0x0022)	O	This command removes an entry from a binding table.
Bind_Register_req (0x0023)	O	This command is unicast by a local device to a primary binding table cache device to register that the local device wants to hold its own binding table.
Replace_Device_req (0x0024)	O	This command is used to replace all existing entries in the binding table of a primary binding table cache related to a specific IEEE/endpoint address with a new entry.
Store_Bkup_Bind_ Entry_req (0x0025)	O	This command is generated by a local primary binding table cache device and sent to a remote backup binding table cache device to request backup storage of a binding table entry.
Remove_Bkup_Bind_ Entry_req (0x0026)	O	This command is generated by a local primary binding table cache device and sent to a remote backup binding table cache device to request removal of a binding table entry.
Backup_Bind_Table_req (0x0027)	O	This command is generated by a local primary binding table cache device and sent to a remote backup binding table cache device to request backup storage of its entire binding table.
Recover_Bind_Table_req (0x0028)	O	This command is generated by a local primary binding table cache device and sent to a remote backup binding table cache device to request a complete restore of its binding table.

Table B.2: (Continued)

Command (*ClusterID*)	Server	Description
Backup_Source_Bind_req (0x0029)	O	This command is generated by a local primary binding table cache device and sent to a remote backup binding table cache device to request backup storage of its source binding table.
Recover_Source_Bind _req (0x002a)	O	This command is generated by a local primary binding table cache device and sent to a remote backup binding table cache device to request complete restore of the source binding table.

Table B.3: Network Management (Client Services) Commands

Command (*ClusterID*)	Server	Description
Mgmt_NWK_Disc_req (0x0030)	O	The local device unicasts this command to a remote device requesting the remote device to perform a scan and identify the existing networks in the vicinity of the local device.
Mgmt_Lqi_req (0x0031)	O	This command requests the neighbor table and associated LQIs of a remote device. The remote device must be a ZigBee coordinator or router.
Mgmt_Rtg_req (0x0032)	O	The local device unicasts this command to a remote device (ZigBee coordinator or router) to request the content of the routing table of the remote device.
Mgmt_Bind_req (0x0033)	O	The local device unicasts this command to a remote device (ZigBee coordinator or router) to request the content of the binding table of the remote device.
Mgmt_Leave_req (0x0034)	O	The local device unicasts this command to a remote device to request the remote device itself or another device to leave the network.
Mgmt_Direct_Join_req (0x0035)	O	With this command, the local device requests the remote device to permit another device to join the network directly. The address of the device that wants to join the network is provided as part of this command.

(Continued)

Table B.3: (Continued)

Command (*ClusterID*)	Server	Description
Mgmt_Permit_Joining_ req (0x0036)	M	This command is broadcast (or unicast) to remote devices requesting them to allow or disallow the association for a specified period of time known as *permit duration*.
Mgmt_Cache_req (0x0037)	O	The local device uses this command to request the list of ZigBee end devices registered with a primary discovery cache device.

Table B.4: Device and Service Discovery (Server Services) Commands

Command (*ClusterID*)	Server	Description
NWK_addr_rsp (0x8000)	M	This is a response to a *NWK_addr_req* command. If the remote device IEEE address matches the IEEE address provided by the *NWK_addr_req*, the remote device will unicast the *NWK_addr_rsp* back with status of success and include the remote device NWK address. The response may optionally include the list of NWK addresses of the remote device associated devices as well.
IEEE_addr_rsp (0x8001)	M	This is a response to an *IEEE_addr_req* command. If the remote device NWK address matches the NWK address provided by the *IEEE_addr_req*, the remote device will unicast the *IEEE_addr_rsp* back with status of success and include the remote device IEEE address. The response may optionally include the list of NWK addresses of the remote device associated devices as well.
Node_Desc_rsp (0x8002)	M	This command provides the node descriptor of the remote device in response to a *Node_Desc_req* command.
Power_Desc_rsp (0x8003)	M	This command provides the power descriptor of the remote device in response to a *Power_Desc_ req* command.
Simple_Desc_rsp (0x8004)	M	This command provides the simple descriptor of the remote device in response to a *Simple_Desc_ req* command.

Table B.4: (Continued)

Command (*ClusterID*)	Server	Description
Active_EP_rsp (0x8005)	M	This command provides the list of active endpoints of the remote device in response to an *Active_EP_req* command.
Match_Desc_rsp (0x8006)	M	The remote device uses this command to provide a list of all its endpoints that match the criteria provided in the *Match_Desc_req* command.
Complex_Desc_rsp (0x8010)	O	This command provides the complex descriptor of the remote device in response to a *Complex_ Desc_req* command.
User_Desc_rsp (0x8011)	O	This command provides the user descriptor of the remote device in response to an *User_Desc_ req* command.
Discovery_Cache_rsp (0x8012)	O	This command is generated with status of success only by a primary discovery cache device in response to a *Discovery_Cache_req*.
User_Desc_conf (0x8014)	O	This command is generated in response to the *User_Desc_set* command. The remote device uses the user descriptor provided by the *User_Desc_set* command to configure the remote device. The result of the configuration is unicast back to the local device using the *User_Desc_conf* command.
System_Server_Discovery_rsp (0x8015)	M	This command is generated in response to a *System_Server_Discovery_req* if the remote device has some of the system server functionalities listed in the *System_Server_Discovery_req*.
Discovery_store_rsp (0x8016)	O	This command is used in response to the *Discovery_store_req*. If the remote device is a primary discovery cache device and has sufficient storage space, it will reply with the status of SUCCESS.
Node_Desc_store_rsp (0x8017)	O	This command is used in response to the *Node_ Desc_store_req*. If the remote device is a primary discovery cache device and has sufficient storage space, it will reply with the status of SUCCESS.
Power_Desc_store_rsp (0x8018)	O	This command is used in response to the *Power_ Desc_store_req*. If the remote device is a primary discovery cache device and has sufficient storage space, it will reply with the status of SUCCESS.

(Continued)

Table B.4: (Continued)

Command (*ClusterID*)	Server	Description
Active_EP_store_rsp (0x8019)	O	This command is used in response to the *Active_EP_store_req*. If the remote device is a primary discovery cache device and has sufficient storage space, it will reply with the status of SUCCESS.
Simple_Desc_store_rsp (0x801a)	O	This command is used in response to the *Simple_Desc_store_req*. If the remote device is a primary discovery cache device and has sufficient storage space, it will reply with the status of SUCCESS.
Remove_node_cache_rsp (0x801b)	O	This command is used in response to the *Remove_node_cache_req*. If the remote device is a primary discovery cache device and contains the discovery information for the device of interest, it will remove the discovery information and reply with the status of SUCCESS.
Find_node_cache_rsp (0x801c)	O	This command is used to notify a local device of the successful discovery of the primary discovery cache device for the given NWK and IEEE addresses.

Table B.5: Binding-related (Server Service) Commands

Command (*ClusterID*)	Server	Description
End_Device_Bind_rsp (0x8020)	O	This command provides the status (SUCCESS, NO_MATCH, etc.) in response to an *End_Device_Bind_req* command.
Bind_rsp (0x8021)	O	This command provides the status (SUCCESS, TABLE_FULL, etc.) in response to a *Bind_req* command.
Unbind_rsp (0x8022)	O	This command provides the status (SUCCESS, NO_ENTRY, etc.) in response to an *Unbind_req* command.
Bind_Register_rsp (0x8023)	O	This command is used to respond to the *Bind_Register_req* command. If the remote device is the primary cache device and contains the local device binding table, it will reply with a status of SUCCESS and include the binding table list in the reply.

Table B.5: (Continued)

Command (*ClusterID*)	Server	Description
Replace_Device_rsp (0x8024)	O	Upon receipt of a *Replace_Device_req* from a local device, the remote device will look for the specified device and if such a device is found, the remote device will replace the old addresses with the new one provided by the *Replace_Device_req*. Then the remote device notifies the local device regarding the status of its request using the *Replace_Device_rsp* command.
Store_Bkup_Bind_Entry_rsp (0x8025)	O	This command provides the status (SUCCESS, TABLE_FULL, etc.) in response to a *Store_Bkup_Bind_Entry_req* command.
Remove_Bkup_Bind_Entry_rsp (0x8026)	O	This command provides the status (SUCCESS, NO_ENTRY, etc.) in response to a *Remove_Bkup_Bind_Entry_req* command.
Backup_Bind_Table_rsp (0x8027)	O	This command provides the status (SUCCESS, TABLE_FULL etc.) in response to a *Backup_Bind_Table_req* command. The response also contains the number of entries in the backup bind table.
Recover_Bind_Table_rsp (0x8028)	O	Upon receipt of a *Recover_Bind_Table_req* from a local device, if the remote device is a backup binding table cache device and contains the specified binding table, will reply using the *Recover_Bind_Table_rsp* and include the binding table in the response.
Backup_Source_Bind_rsp (0x8029)	O	This command provides the status (SUCCESS, TABLE_FULL, etc.) in response to a *Backup_Source_Bind_req* command.
Recover_Source_Bind_rsp (0x802a)	O	Upon receive of a *Recover_Source_Bind_req* from a local device, if the remote device is a backup binding table cache device and contains the specified source binding table, will reply using the *Recover_Source_Bind_rsp* and include the source binding table in the response.

Table B.6: Network Management (Server Services) Commands

Command (*ClusterID*)	Server	Description
Mgmt_NWK_Disc_rsp (0x8030)	O	This command is a response to a *Mgmt_NWK_Disc_ req* and provides the list of networks found using a network scan. The characteristic of each network, such as logical channel and 64-bit PAN identifier, are also provided.
Mgmt_Lqi_rsp (0x8031)	O	This command is generated in response to the *Mgmt_ Lqi_req* command and provides the neighbor table and associated LQIs of the remote device.
Mgmt_Rtg_rsp (0x8032)	O	This command is generated in response to the *Mgmt_ Rtg_req* command and provides the content of the routing table of the remote device.
Mgmt_Bind_rsp (0x8033)	O	This command is generated in response to the *Mgmt_ Bind_req* command and provides the content of the binding table of the remote device.
Mgmt_Leave_rsp (0x8034)	O	This command is generated in response to a *Mgmt_ Leave_req* and provides the status of the remote device attempt to make a remote device leave the network.
Mgmt_Direct_ Join_rsp (0x8035)	O	This command provides the status of direct join attempt (requested by a *Mgmt_Direct_ join_req*) back to the local device that initiated the request.
Mgmt_Permit_Joining_rsp (0x8036)	M	This command provides the status (SUCCESS, etc.) in response to a *Mgmt_Permit_Joining_req* command.
Mgmt_Cache_rsp (0x8037)	O	This command is generated in response to a *Mgmt_ Cache_req* and provides the list of ZigBee end devices registered with a primary discovery cache device.

DSSS Symbol-to-Chip Mapping Tables

Table C.1: DSSS Symbol-to-Chip Mapping (OQPSK)

Data Symbol (b0,b1,b2,b3)	Chip Value (c0,c1,...,c31)
0 0 0 0	1 1 0 1 1 0 0 1 1 1 0 0 0 0 1 1 0 1 0 1 0 0 1 0 0 0 1 0 1 1 1 0
1 0 0 0	1 1 1 0 1 1 0 1 1 0 0 1 1 1 0 0 0 0 1 1 0 1 0 1 0 0 1 0 0 0 1 0
0 1 0 0	0 0 1 0 1 1 1 0 1 1 0 1 1 0 0 1 1 1 0 0 0 0 1 1 0 1 0 1 0 0 1 0
1 1 0 0	0 0 1 0 0 0 1 0 1 1 1 0 1 1 0 1 1 0 0 1 1 1 0 0 0 0 1 1 0 1 0 1
0 0 1 0	0 1 0 1 0 0 1 0 0 0 1 0 1 1 1 0 1 1 0 1 1 0 0 1 1 1 0 0 0 0 1 1
1 0 1 0	0 0 1 1 0 1 0 1 0 0 1 0 0 0 1 0 1 1 1 0 1 1 0 1 1 0 0 1 1 1 0 0
0 1 1 0	1 1 0 0 0 0 1 1 0 1 0 1 0 0 1 0 0 0 1 0 1 1 1 0 1 1 0 1 1 0 0 1
1 1 1 0	1 0 0 1 1 1 0 0 0 0 1 1 0 1 0 1 0 0 1 0 0 0 1 0 1 1 1 0 1 1 0 1
0 0 0 1	1 0 0 0 1 1 0 0 1 0 0 1 0 1 1 0 0 0 0 0 1 1 1 0 1 1 1 1 1 0 1 1
1 0 0 1	1 0 1 1 1 0 0 0 1 1 0 0 1 0 0 1 0 1 1 0 0 0 0 0 0 1 1 1 0 1 1 1
0 1 0 1	0 1 1 1 1 0 1 1 1 0 0 0 1 1 0 0 1 0 0 1 0 1 1 0 0 0 0 0 0 1 1 1
1 1 0 1	0 1 1 1 0 1 1 1 1 0 1 1 1 0 0 0 1 1 0 0 1 0 0 1 0 1 1 0 0 0 0 0
0 0 1 1	0 0 0 0 0 1 1 1 0 1 1 1 1 0 1 1 1 0 0 0 1 1 0 0 1 0 0 1 0 1 1 0
1 0 1 1	0 1 1 0 0 0 0 0 0 1 1 1 0 1 1 1 1 0 1 1 1 0 0 0 1 1 0 0 1 0 0 1
0 1 1 1	1 0 0 1 0 1 1 0 0 0 0 0 0 1 1 1 0 1 1 1 1 0 1 1 1 0 0 0 1 1 0 0
1 1 1 1	1 1 0 0 1 0 0 1 0 1 1 0 0 0 0 0 0 1 1 1 0 1 1 1 1 0 1 1 1 0 0 0

Table C.2: DSSS Symbol-to-Chip Mapping (BPSK)

Input Bit	Chip Value (c0,c1,...,c14)
0	1 1 1 1 0 1 0 1 1 0 0 1 0 0 0
1	0 0 0 0 1 0 1 0 0 1 1 0 1 1 1

ZigBee-Pro/2007

The original version of the ZigBee standard was released in 2004. The second release of the ZigBee standard was in 2006 and is referred to as ZigBee-2006. The third version came out in 2007. When the updated ZigBee specifications were released in 2007, the stack profile was renamed to ZigBee-Pro. Some of the optional features in the ZigBee-2006 have become mandatory in ZigBee-Pro/2007.

If you implement ZigBee-2006 or ZigBee-Pro/2007 on your device, the device is not required to support backward compatibility with ZigBee-2004 devices. If a device implements the ZigBee-Pro/2007, however, the device must be backward compatible to ZigBee-2006. A ZigBee-2006 device can join a ZigBee-Pro network, but the ZigBee-2006 device can only act as an end device in a ZigBee-Pro network. In other words, a ZigBee-2006 node cannot route the messages or accept children in a ZigBee-Pro network.

For some applications, implementing ZigBee-2006 can be sufficient and there might not be a need to upgrade the network to ZigBee-Pro/2007. The theoretical network size in ZigBee-2006 and ZigBee-Pro/2007 are the same, but ZigBee-Pro has additional features that make it a more suitable choice for large size networks (e.g., a network with thousands of nodes) compared to ZigBee-2006. ZigBee-Pro allows transmission of large messages using fragmentation and reassembly mechanism. The frequency agility is a mandatory feature in ZigBee-Pro/2007; it is optional in ZigBee-2006.

In this book, the information regarding ZigBee-Pro is provided along with ZigBee-2006. This appendix summarizes some of the differences between ZigBee-Pro/2007 and ZigBee-2006.

D.1   Frequency Agility

In a ZigBee network, the ZigBee coordinator selects a frequency channel for the entire network. If the network operates in the presence of strong interferers, it might be necessary to change the frequency channel for the entire network. Even though in a ZigBee-2006 network it is possible to change the frequency channel whenever needed, ZigBee-Pro includes additional features that facilitate switching the frequency channel. For example, in a ZigBee-Pro network, there is a dedicated device called the *network channel manager* that receives reports of interference issues from other nodes in the network. If a router in the network faces frequent transmission errors, it will notify the network channel manager about a persistent channel access issue. When the network channel manager receives sufficient reports of interferers in the current frequency channel from various nodes, the network channel manager selects a new channel for the network based on the reports of channel energy measurements received from several nodes in the network. After selecting a new frequency channel, the network channel manager informs all the nodes in the network about the new frequency channel using a ZDO command. This is referred to as *frequency agility*. The frequency agility of a ZigBee-Pro network improves its coexistence performance.

D.2   Address Allocation

Address allocation, discussed in Chapter 3, was based on ZigBee-2006. In ZigBee-Pro, in addition to distributed address allocation, stochastic address allocation is supported. In stochastic address allocation mode, a device or its parent picks a random address for the device. The address can be any random address other than addresses already listed in the NWK Information Base (NIB). In ZigBee-Pro, in contrast to ZigBee-2006, a device can keep its randomly assigned address even if it rejoins a network, as long as there is no address conflict with any other device on the network. In the event of an address conflict, the device can pick a new random address or rejoin the network to obtain a new random address. In ZigBee-Pro, ZigBee routers may have up to 20 children, but only six of these children can act as routers. The rest of the children must be end devices.

D.3   Security

ZigBee-Pro supports additional security features that are not available in ZigBee-2006. For example, in ZigBee-2006, there are nine APS security commands (discussed in Chapter 3). In ZigBee-Pro there are five more commands, in addition to the commands supported in ZigBee-2006. One of these additional commands (the *Tunnel* command) allows a device to send a command to a device that does not have the current network key. The rest of the additional commands are used for entity authentication, which allows two devices to mutually authenticate each other. These two devices must share a common security key. Each device creates a 16-octet random string called *random challenge* and sends this challenge to the other device. These 16-octet challenges are used as part of Mutual Symmetric-Key Entity Authentication Scheme to verify the authenticity of two devices concurrently. Entity authentication is not supported in ZigBee-2006.

In ZigBee-2006, the trust center can operate in commercial mode or residential mode. Commercial mode maintains a higher security level than residential mode. For example, in commercial mode, the trust center must maintain a list of devices, master keys, link keys, and network keys. But in residential mode, only the network key must be maintained. ZigBee-Pro uses the same concept but renames the commercial and residential modes of operation to high security and standard security modes accordingly. In ZigBee-2006, the trust center is assumed to be located on the ZigBee coordinator. The trust center in ZigBee-Pro can be on any device.

D.4   Routing

Figure D.1 shows a routing example from a source device (device 1) to a destination device (device 8). The link cost for sending a message from device 1 to device 2 is represented by $C\{D_1,D_2\}$. Similarly, the link cost for sending a message from device 2 to device 1 can be shown by $C\{D_2,D_1\}$. The maximum output powers and receiver sensitivities of device 1 and device 2 might not be the same. The link cost, in general, depends on the direction of the message; therefore, $C\{D_1,D_2\}$ and $C\{D_2,D_1\}$ might not be equal. This means that if a route is the optimum choice to carry a message in one direction, the same route might not be the best option to carry the message in the opposite direction.

In ZigBee-2006, when a route is established from a source device to a destination device, the same route will be used to carry the messages back from the destination device to the source device. This is known as *symmetric routing*. In *asymmetric routing*, in contrast, the network carries the message from the source device to the destination device using a particular route, but a different route is used to carry a message back to the source device from the destination device (Figure D.1). One of the main routing features added to ZigBee-Pro/2007 is asymmetric routing capability. Asymmetric routing increases the network reliability and robustness because it allows picking the best path in both directions.

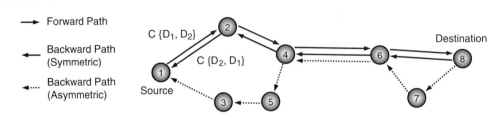

Figure D.1: Asymmetric Routing

In the ZigBee-2006, the NWK layer *route request* command can only be used to request a unicast or a multicast route (see Figure 3.44 in Chapter 3). In ZigBee-Pro, the *route request* command can support many-to-one route requests as well. The sender of the many-to-one route request is not required to support a route record table.

ZigBee-Pro also replaces the NWK layer *route error* command with the *network status* command. The *route error* commands contains an error code that determines the type of error. ZigBee-Pro renames the error code to *status code*. The status code in the network status command of ZigBee-Pro contains all the error codes in ZigBee-2006 plus additional codes to report incidents such as an update in the PAN identifier.

The NWK layer in the ZigBee-Pro has three additional commands that are not available in ZigBee-2006. The first one is the *link status* command. Each router can use the *link status* command to communicate its link cost to other neighboring routers. The second command is the *network report* command, which is used to report network events such as PAN ID conflict and channel condition to a designated device in the *nwkManagerAddr*. The last command is the *network update* command. This command is used by a designated device (identified by *nwkManagerAddr* attribute) to broadcast configuration changes.

In ZigBee-2006, a NWK frame may require *force route discovery*. This means that a route discovery will be initiated for transmitting this frame, even if there is already a route established to the destination. In ZigBee-Pro, a NWK frame can no longer request force route discovery, and the force route discovery option is removed from the discover route subfield of the NWK frame.

Some of the routers in the network might be battery powered. ZigBee-Pro distinguishes the battery-powered routers and marks them as low-power routers (LPRs). When broadcasting a message, you have the option to broadcast the message to all routers or only to low-power routers. The additional features available in ZigBee-Pro can be used to improve the routing efficiency.

D.5   Fragmentation and Reassembly

In ZigBee-Pro, if the APS Service Data Unit (APSDU) is larger than the maximum size of the payload of a single NWK frame, it is possible to break the APSDU to smaller sizes that fit the NWK payload before transmission. The receiver understands that the received messages contain fragmented portions of the APSDU and will reconstruct the APSDU to its original form. This is referred to as *fragmentation and reassembly*. ZigBee-2006 does not support fragmentation and reassembly. If fragmentation and reassembly are not supported, the APSDU is limited to the maximum size of the payload of a single NWK frame.

References

[1] ZigBee Specifications (2006), available from www.zigbee.org.
[2] ZigBee Specifications (2007), available from www.zigbee.org.

Transceiver Building Blocks

This appendix is a simple tutorial that reviews some of the main building blocks inside a typical low-power and low-cost short-range wireless networking transceiver. This appendix, however, is not a prerequisite for understanding chapter 4 of this book.

E.1   Introduction

The transceiver architecture and topology of individual blocks inside a transceiver are determined based on the application scenario. The transceivers developed for IEEE 802.15.4 standard must be very low cost and consume very little power. The IEEE 802.15.4 performance requirements, such as error vector magnitude (EVM) and receiver sensitivity, have been relaxed by the standard developers to allow the manufacturers select simple and low cost architectures for their transceivers. There are several textbooks that provide basic understanding of transceiver topologies and design of individual blocks inside a transceiver including [12]. For simplicity, the transceiver topology selected for this appendix consists of a direct down conversion receiver and a direct up conversion transmitter, shown in Figure E.1. In this appendix, first the components in the receiver chain are reviewed. Then an overview of the transmitter building blocks is provided. This appendix also discusses some of popular microcontrollers, interfaces, and package types used in low-power short-range wireless networking transceivers.

E.2   Receiver Chain Building Blocks

In a direct conversion receiver, the local oscillator (LO) frequency is the same as the input RF signal carrier frequency. When the RF signal is multiplied by the LO signal, the input signal is down converted directly to the baseband. The direct conversion receiver

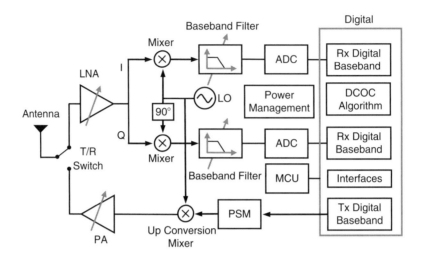

Figure E.1: Simplified Diagram of a Direct Conversion Transceiver

(DCR) can be highly integrated and requires only a few external components. Reducing the number of external components reduces the total cost and size of the transceiver. The main challenges in a DCR are the DC offset and flicker noise. The sources of the DC offset and basic solutions for DC offset corrections are provided in section E.2.4. The flicker noise is a low frequency noise that becomes stronger as the frequency is reduced. The flicker noise is mainly due to active devices and increasing the size of these devices helps reducing the flicker noise. Increasing the size of a device increases the parasitic capacitances associated with the device. The parasitic capacitances can degrade the performance of an RF block. Therefore, increasing the devices size may not be an option in an RF block such as a low noise amplifier (LNA). In low frequency blocks such as baseband filter, the additional parasitic capacitances are not significant contributors to the performance of the block. In DCR, the flicker noise of the LNA is removed by an AC coupling capacitor between the LNA and mixer.

One of the basic properties of a signal is its signal-to-noise ratio (SNR). The SNR is simply the ratio of the signal power to the noise power within the frequency band of interest. Every time a signal passes through a block, the signal SNR will be degraded because the block will add its noise to the signal. The noise figure (NF) is the ratio of the input signal SNR to the output signal SNR. An ideal noiseless block has NF of 0 dB.

A receiver must meet several performance metrics. The dynamic range, for example, is the ratio of the smallest detectable input signal power to the largest input signal

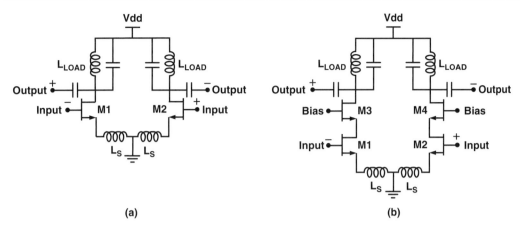

Figure E.2: Simplified Diagram of (a) Differential LNA (b) Cascode LNA

power that the receiver can successfully detect. In IEEE 802.15.4 standard, the receiver sensitivity is defined as the lowest received signal power that yields Packet Error Rate (PER) of less than 1%. The IEEE 802.15.4 requires only $-85\,$dBm of sensitivity for operations in 2.4 GHz ISM band. In 868/915 MHz band, if the BPSK modulation is used, the required sensitivity is $-92\,$dBm. The optional modes of operation in 868/915 MHz band (using ASK and OQPSK modulation) must meet $-85\,$dBm of sensitivity. The receiver sensitivity can be determined from the receiver noise figure [12]. In this section, some of the main blocks inside a typical direct conversion receiver are reviewed and examples of design trade-offs are provided.

E.2.1 LNA

The LNA gain and NF are the most important contributors to the receiver sensitivity. The antenna impedance is normally a known fixed value. In order to maximize LNA gain, the LNA input impedance must be the conjugate match of the antenna impedance. However, the input impedance that is optimum for LNA gain may not result in the optimum NF for the LNA. Therefore, the LNA input impedance is always selected by a trade-off between the LNA gain and LNA NF performances. In a direct conversion receiver, there is no image rejection filter between the LNA and the mixer.

The LNA design starts with selection of the LNA topology. A simple topology for LNA is a differential amplifier with inductive loads, shown in Figure E.2a. The LNA load can be either a resistor or an inductor. Using inductors instead of resistors for LNA load improves the LNA linearity and NF performances, but the area occupied by the inductors may be too large for a low cost transceiver. Considering that IEEE 802.15.4 receiver sensitivity requirement is relatively easy to meet, it is possible to use resistors for the LNA load and still meet IEEE 802.15.4 requirements. The inductors L_{S1} and L_{S2} help improve LNA linearity. In majority of IEEE 802.15.4 modes of operations, the modulated signal has a constant envelope and the signal is not very sensitive the LNA nonlinearities. Therefore, improving LNA linearity using inductors L_{S1} and L_{S2} may not be necessary for a low cost IEEE 802.15.4 transceiver.

The Figure E.2b shows a cascode LNA topology. The additional devices (M3 and M4) help improve the LNA stability and reverse isolation at the cost of degrading LNA NF performance. Reverse isolation is particularly important in direct conversion receivers as the LO signal is at the same frequency as the RF and the leakage of the LO signal to the LNA input must be minimized.

The main sources of the nonlinearities in an LNA are the active devices. In CMOS, the drain current is approximately proportional to the square of the input AC voltage. The odd order nonlinearities can degrade the shape of the power spectral density of the desired signal. The even order nonlinearities, however, are less critical in an LNA because the even order nonlinearities create undesired signals either around DC or at very high frequencies. The LNA is connected to the mixer by a capacitor, which removes all of the undesired spectral contents around DC before they reach the mixer. The undesired spectral contents at frequencies much higher than the desired channel frequency will be filtered out and are less likely to affect receiver performance.

E.2.2 Mixer

Figure E.3a is a simplified diagram of a Gilbert cell, which is one of the commonly used active mixer topologies. The differential RF signal enters the mixer through M1 and M2. The devices M1 and M2 along with resistors R1 and R2 form a differential amplifier. The differential LO signal turns on and turns off the switching devices (M3 to M6) with LO frequency and causes the RF signal to be down converted to the baseband. In mixer design, the goal is to minimize the NF, area, current consumption, and flicker noise, while maximizing the gain, linearity, and port-to-port isolation. Poor isolation between the ports causes undesired DC offset. The junction capacitances of the devices, for example, make undesired paths between the LO, RF, and baseband ports of a mixer. If the LO signal, which is normally much stronger than the RF input, leaks to the RF port, it will be self mixed with the LO signal. One of the results of multiplication of a single tone with another

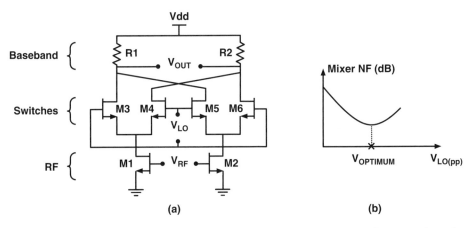

Figure E.3: (a) Gilbert Cell Mixer (b) Mixer NF Versus LO Peak-to-Peak Voltage

single tone with the same frequency is a DC term. The undesired DC offset at the output of the mixer can be reduced by a DC offset correction mechanism, discussed in section E.2.4.

Figure E.3b shows that the mixer noise figure is a function of the maximum peak-to-peak voltage of the LO signal ($V_{LO(pp)}$). The reason for mixer NF degradation when the LO signal is too strong is that by increasing the LO voltage, the LO signal becomes almost a square shaped signal. A square-shaped signal contains additional harmonics that are not present in a pure sinusoid signal. These additional harmonics down convert additional noise to the baseband and degrade the desired signal NF. On the other hand, if the LO signal is not strong enough, the switches M3 to M6 will not turn on and turn off fast enough and there will be moments in time that all four switching devices are conducting. This concurrent conduction of all the switching devices provides a path for the noise to go directly from devices M1 and M2 to the resistors R1 and R1 without any frequency conversion and consequently degrade the mixer NF.

E.2.3 Base-Band Filter and AGC

After the signal is down converted by the mixer, the signal must be amplified and filtered before it is delivered to the ADC block. As it is shown in Figure E.4, the filtering and amplifying are performed in multiple stages. For simplicity, Figure E.4 shows only one of the receiver channels. The filter order and bandwidth is determined by the desired signal bandwidth, the adjacent and alternate channel requirements, and the ADC sampling rate (if applicable).

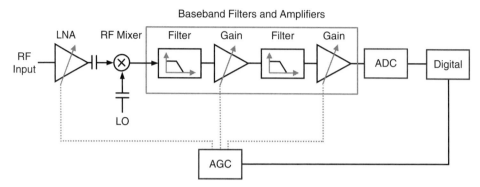

Figure E.4: Baseband Filter and Variable Gain Amplifier Stages (Only I Channel is Shown Here)

An IEEE 802.15.4 compliant receiver must meet adjacent and alternate channels rejection requirements specified in the IEEE 802.15.4 standard. After the filtering, the signals in the adjacent and alternate channels become weaker than the desired signal. This would allow the digital baseband to demodulate the desired signal successfully in presence of adjacent and alternate channel interferences.

An oversampling ADC is an ADC that samples the input signal at a much higher rate than the Nyquist rate. If an oversampling ADC is used in the receiver chain, the baseband filter must act as an anti-aliasing filter for the ADC block. The baseband filtering removes the noise outside the frequency band of interest and improves the desired signal SNR.

The received RF signal power may vary significantly over time and the receiver chain amplifiers must be able to change their gains based on the received signal strength. The role of the automatic gain controller (AGC) in a receiver is to measure the received signal strength and change the gain of the amplifying stages dynamically to make sure the signal delivered to the ADC is within the dynamic range of the ADC. An IEEE 802.15.4 must be capable of measuring the received signal strength (RSS) and generating received signal strength indicator (RSSI). Therefore, AGC can use the available RSS information and adjust the receiver gain stages accordingly. If the LNA gain is controllable, the AGC will have the option to reduce the LNA gain whenever the received signal is very strong. This would avoid saturating the baseband filter and ADC blocks.

The RF signal that enters the receiver chain through the LNA not only contains the desired signal, but also may contain adjacent and alternate channel interfering signals that are stronger than the desired signal. Although the baseband filter will attenuate the signals outside the desired channel, the LNA and mixer amplify all of the signals within ISM band equally. Therefore, the gain control mechanism needs to measure the signal energy outside the desired channel as well as within the desired channel and include them in gain control mechanism to avoid saturating the receiver chain blocks. One of the baseband filter design challenges in direct conversion receivers is the undesired DC offset. The sources of DC offset and DC offset correction mechanisms are discussed in section E.2.4.

The frequency bandwidth of each stage of the filter is typically depends on some resistors and capacitors. The value of each capacitor or resistor may vary up to 50% due to process variations. Consequently, the bandwidth of filter may vary up to 50%. This large variation is unacceptable in many applications and several methods have been developed to reduce baseband filter bandwidth uncertainty due to process variations. One of these methods is the pole-tracking mechanism. In this approach, an external capacitor is used as an accurate reference to determine the variations of the on-chip capacitor values. The value of the external capacitor is independent of the process variations and comparing the value of the external capacitor with an on-chip capacitor provides an estimate of the capacitance variations for all capacitors of the same type that are co-located on the same die. Assuming the values of the capacitors or resistors used in a baseband filter are adjustable, the pole-tracking method can be used to adjust the values of capacitors and resistors of the baseband filter accordingly. Using this method, it is possible to reduce the variations of the baseband filter bandwidth to less than 10%.

If the minimum gain of the baseband filter is zero dB, the required maximum gain of the baseband filter can be determined from the following equation:

$$G_{BB(MAX)} = D_R - D_{ADC} - \Delta G_{LNA} \,(dB) \qquad (E.1)$$

where $G_{BB(MAX)}$ is the maximum gain of the baseband filter, D_R is the required dynamic range of the receiver, D_{ADC} is the dynamic range of the ADC, and ΔG_{LNA} is the difference between the maximum gain of the LNA and minimum gain of the LNA. For example, if the required dynamic range of the receiver is 110 dB, the ADC dynamic range is 55 dB, and the difference between LNA high gain and low gain is 15 dB, then the baseband filter maximum gain must be at least 40 dB.

E.2.4 DC Offset Correction

The undesired DC offset is generated from several sources. Self mixing of the signals in the receiver path, for instance, can create a DC offset. Figure E.5a shows that the signal generated by the local oscillator (LO) can leak to the LNA because of insufficient reverse

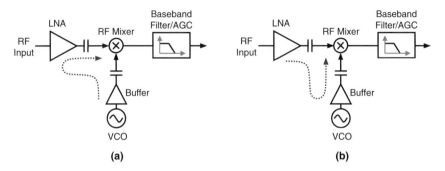

Figure E.5: DC Offset Due to (a) Local Oscillator Self Mixing (b) RF Signal Self Mixing

isolation of the LNA or presence of parasitic capacitances between LNA and LO. This leaked signal will reach the RF mixer along with the desired RF input signal. This leaked signal has the same frequency as the LO signal and when multiplied by the LO will create a DC offset at the output of the mixer. In a direct-conversion receiver, there is no AC coupling between the RF mixer and the baseband filter and therefore, the DC offset caused by LO leakage will reach the baseband filter along with the desired signal. If the LNA gain is not variable, the DC offset caused by LO leakage is a constant value and correcting this DC offset is relatively simple.

Figure E.5b shows another source of DC offset in a receiver. The RF signal is leaking to the LO and is multiplied by itself at the RF mixer. The result is a DC offset at the input of the baseband filter. The DC offset caused by the RF signal leakage is challenging to remove because the value of this DC offset is varying based on the RF signal strength. In order to address this issue, several dynamic DC offset correction mechanisms have been developed that can remove time varying DC offsets.

A portion of the DC offset can be the result of device mismatches. A designer may use two identical devices in his block, but after fabrication, the performance of these two devices may not be exactly the same because of unavoidable variation in the process. Another source of DC offset is even order nonlinearities of the baseband block. Even order nonlinearities act similarly to self mixing mechanisms and create DC offset.

Figure E.6a shows a simple DC offset correction (DCOC) method. In this approach, an analog feedback loop measures the DC offset at the output of the baseband filter and adjusts the DC operating point at the input of the baseband filter accordingly. This simple

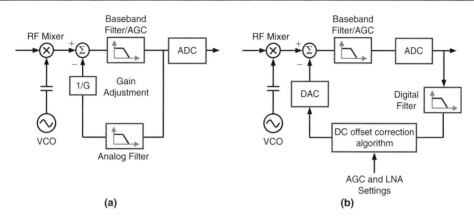

Figure E.6: DC Offset Cancellation Using (a) Analog Feedback (b) Digital Feedback

feedback loop not only eliminates the DC offset, but also removes a small portion of the desired signal. This is because the feedback signal contains a portion of the desired signal, proportional to the bandwidth of the analog filter in the feedback loop. Removing a portion of the desired signal can degrade the BER performance of the receiver. As a result, the bandwidth of the analog filter in the feedback loop must be significantly smaller than the bandwidth of the baseband filter. However, reducing the bandwidth of this analog filter in the feedback loop will slow down the DC offset correction. Generally, it is desired to correct the DC offset as fast as possible because a slow DCOC loop may not be able to adjust the DC offset in time to avoid receiver BER performance degradation. Therefore, the bandwidth of the analog filter in the feedback loop is selected based on a trade-off between feedback loop settling time and the signal PSD degradation. The total area available for the analog filter can also play a role in selecting its bandwidth because the lower its bandwidth is, the larger the size of analog filter will be.

Figure E.6b shows an alternative DCOC method, where the DC offset is measured at the output of the ADC. The filter that isolates the DC offset is a digital filter. Since the signal is digital, variety of algorithms can be used to correct the DC offset. For example, the DCOC algorithm may have two modes of operation. In the first mode (learning mode), the algorithm calculates the value of the DC feedback for different AGC and LNA settings, while there is no RF signal present. During the live operation, the algorithm can use the previously stored values as its initial guess for the proper DC feedback and then calculates the precise value of the DC feedback. In this way, this DCOC algorithm can correct the

DC offset faster than algorithms that do not have any learning period. After calculating the proper amount of the feedback, a digital to analog converter (DAC) is required to create the analog DC voltage and subtract it from the baseband filter input. Since the bandwidth of the filter used in the feedback loop is more than zero, the DC offset correction not only reduces the DC offset, but also decreases the amount of the flicker noise.

The DC offset correction methods shown in Figure E.6b can remove majority of the DC offset in the receiver chain. However, even after the DCOC algorithm reaches its settling point, a small DC offset may remain in the signal delivered to the digital baseband. The reminder of the DC offset can be removed by the digital baseband.

E.2.5   Analog to Digital Converter

The next stage after the analog baseband filter is the analog to digital converter (ADC). An ADC samples the analog signal and assigns a digital value to each sample. An introduction to analog to digital converters and their performance metrics is provided in section 4.12 of chapter 4.

E.2.6   Receiver Digital Baseband

After the signal is converted to digital by the ADC, it is delivered to the receiver digital baseband. The receiver digital baseband is responsible to recover and demodulate the data. The demodulator is one of the main building blocks in a receiver digital baseband. An OQPSK demodulator receives the in-phase (I) and quadrature (Q) samples of the signal and makes a decision on the corresponding symbols. There are different methods to implement a demodulator and selecting the demodulation method is a tradeoff between the receiver sensitivity, implementation complexity, and sensitivity to the received signal phase and frequency errors. Coherent detection, for example, assumes availability of estimated channel characteristics. The channel characteristic is determined by transmitting training signals from the transmitter to receiver. The coherent detection is complex and is not suitable for low-cost IEEE 802.15.4 receivers.

Noncoherent detection is a simple alternative to coherent detection and does not require the channel information. In noncoherent detection, instead of measuring the signal absolute phase, the phase change from one symbol to the next symbol is measured to determine the received data. This is referred to as differential symbol detection. Differential symbol detection eliminates the effect of any constant phase error because only phase difference between two consecutive symbols is taken into account, instead of absolute phase value of each symbol. Noncoherent detection method simplifies the receiver implementation at the cost of degrading the receiver sensitivity in additive white Gaussian noise (AWGN) channels.

The operating frequency of the transmitter device and the receiver device may not be exactly the same and the receiver device must be able to tolerate some level of frequency error in a received signal. An IEEE 802.15.4 transmitter is allowed to have ± 40 ppm of frequency error. One of the methods to improve the frequency offset tolerance of a receiver is using differential chip detection (DCD). The differential chip detection is similar to the deferential symbol detection, but it measures the difference between two consecutive chips instead of two consecutive symbols. The DCD method degrades the receiver sensitivity level in AWGN channels more than noncoherent detection method, but enhances the frequency offset tolerance of the receiver.

E.3 Transmitter Chain Building Blocks

The function of a transmitter is modulating the data, up converting the baseband data to the carrier frequency, amplifying the signal, and finally transmitting the modulated signal. A transmitter may up-convert the baseband data to the carrier in two steps using an intermediate frequency (IF). Proper selection of the IF frequency can help reducing the pulling issue in a transmitter. The pulling issue is discussed in section E.4. If a transmitter up converts the baseband signal directly to the carrier frequency, the architecture is referred to as the direct-conversion. The direct-conversion transmitters are generally simple to implement, but can be highly susceptible to the pulling issue particularly if the LO is generated by an LC-VCO. The direct-conversion transmitters are suitable for low-cost ISM band transmitters because of their simplicity. This section reviews the building blocks in a direct-conversion transmitter.

E.3.1 PSM

The Phase Shift Modulator (PSM) block in the transmitter chain converts the stream of bits into modulated signals. The modulated signal generated by the PSM block will be amplified by the Power Amplifier (PA) block before transmission.

Consider the Offset QPSK (OQPSK) modulation constellation points in Figure E.7a. In OQPSK, every 2 bits of data is mapped into a specific signal phase. The PSM block usually adds intermediate phase shifts to improve the shape of the Power Spectral Density (PSD) of the output signal. If the PSM block does not add any intermediate phase shifts, the phase shift per quadrant is equal to one. One quadrant is equal to one quarter of the

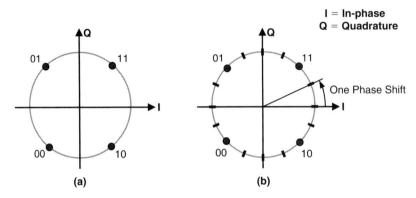

Figure E.7: (a) OQPSK Modulation Constellation Points (b) PSM Intermediate Phase Shifts (4 Shifts per Quadrant in this Example)

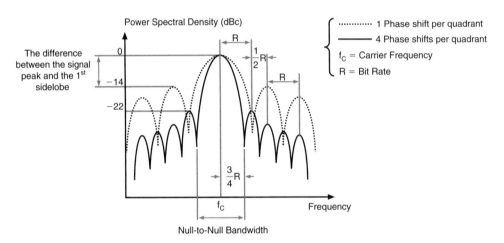

Figure E.8: The OQPSK Power Spectral Densities (Prior to Filtering) for 1 Shift-per-Quadrant and 4 Shifts-per-Quadrant

circle shown in Figure E.7a. In OQPSK, one quadrant is equal to period of one bit. In Figure E.7b, each intermediate phase shift is 22.5 degrees; therefore, there are 4 phase shifts per quadrant.

The effect of the intermediate phase shifts on the output signal PSD is shown in Figure E.8. Both Power Spectral Densities in Figure E.8 are for OQPSK modulated signals with the same bit rate and carrier frequency. The bit rate is shown by "R" in Figure E.8. For example, in IEEE 802.15.4 2.4 GHz mode of operation, the bit rate (R) is equal to 2 Mbps.

From Figure E.8 it can be concluded that by increasing the number of phase shifts per quadrant, the signal energy is more concentrated around the carrier frequency. In the OQPSK signal with 1 phase-shift per quadrant, the first sidelobe is at -14 dBc, while in the OQPSK signal with 4 phase-shifts per quadrant, the first sidelobe is improved to -22 dBc. Increasing the number of phase shifts per quadrant to infinity can only improve the first sidelobe to -23 dBc. Therefore, implementing the PSM block with 4 phase-shifts per quadrant is adequate for an IEEE 802.15.4 transceiver.

The signal bandwidth can be defined as the frequency interval between the nulls on the PSD graph. This is known as null-to-null bandwidth of a signal. Figure E.8 shows that increasing the number of phase-shifts per quadrant will reduce the null-to-null bandwidth of the signal.

The Error Vector Magnitude (EVM) is an indication of the modulation accuracy. The EVM is discussed in chapter 4. The EVM of a signal generated by the PSM block depends on phase noise of the VCO and undesired variations of the signal phase and amplitude in PSM block. Section E.4 provides an approximate way of calculating the relationship between the VCO phase noise and the signal EVM. Compared to other wireless networking standards, the IEEE 802.15.4 has a relaxed EVM requirement. A transmitter EVM must be less than 35% to pass EVM requirement of the IEEE 802.15.4 standard.

There is a relationship between the phase and amplitude error of the signal generated by the PSM block and the level of the first sidelobe (dBc) of the generated signal Power Spectral Density [17]. The higher the phase and amplitude errors are, the higher the level of the first sidelobe will be. For example, 5.73 degrees of phase error will raise the first sidelobe level of an OQPSK signal about 1 dB. The change in the sidelobe level is only apparent before the PSM output signal is filtered or amplified by a nonlinear PA.

The PSM uses a reference clock frequency. The spurious generated by the PSM are normally located at the integer multiples of the PSM clock frequency. For example, if the clock frequency of the PSM block is 16 MHz, then there is a chance that the spurious signals appear at ± 16 MHz and ± 32 MHz of the signal carrier frequency. If these spurious signals are not suppressed before the signal is transmitted by the Power Amplifier (PA), the spurious emissions may cause interference to other IEEE 802.15.4 channels. The PSM block may also create spurious signals at integer multiples of the signal carrier frequency. For instance, if the signal carrier frequency is 2450 MHz, then the spurious signals may be present at 4900 MHz and 7350 MHz. These spurs are located outside the frequency band of interest and can be filtered out easily.

E.3.2 PA

The Power Amplifier (PA) is the last stage of the transmitter chain. The PA amplifies the signal and delivers it to the antenna. Typically, the commercially available transceivers for IEEE 802.15.4 networking come with an integrated PA on the same silicon die as the rest of the transceiver blocks. The output power of an integrated PA is around ± 3 dBm, which is far less than most of the standalone ISM band power amplifiers. A standalone PA can be implemented using a high performance process technology, but the integrated PAs are mostly fabricated using standard CMOS process to reduce the transceiver cost. This section discusses only the integrated power amplifiers.

One simple way to review basic characteristics of power amplifiers is using an example. Figure E.9 shows a single-ended PA with inductive load. The integrated power amplifiers are normally differential, but a single-ended is used in this example for simplicity. The characteristics discussed for this single ended power amplifier are applicable to differential ones as well. The transistor M1 in Figure E.9 converts the input voltage to current. The input voltage V_{in} is the modulated signal (at RF frequency) that will be transmitted. The gate oxides in standard CMOS processes have low break down voltages. The inductor L causes the V_{out} swing above and below V_{dd} voltage. If the transistor M2 was not present, increasing V_{out} voltage above V_{dd} could cause irreversible damages to M1. The transistor M2 not only protects M1 gate oxide, but also improves the reverse isolation of the PA.

In Figure E.9, the antenna, external matching components (if present), the Transmit/ Receive (T/R) switch, and the decoupling capacitor are modeled by an equivalent impedance (R + jX). In chapter 5, the equivalent impedance of an antenna and the

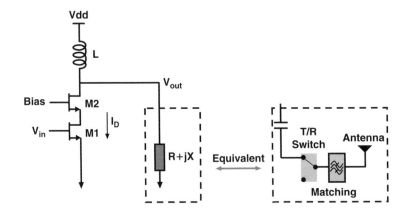

Figure E.9: Simplified Diagram of a Power Amplifier and its Equivalent Load

power transfer efficiency between the antenna and the transceiver are discussed. Proper matching between the PA and the antenna is necessary to ensure maximum power will be radiated from the antenna.

A significant portion of the current consumption in the transmitter chain can be due to the power amplifier. Therefore, optimizing the performance of a PA is essential to reduce the overall current consumption of the transmitter chain. The PA efficiency (E_{PA}) is defined as the ratio of the power delivered by the PA to the antenna to the DC power consumed by the PA:

$$E_{PA} = \frac{P_{out\ (rms)}}{P_{DC}} \times 100(\%) \tag{E.2}$$

For example, a PA that delivers 0 dBm output power, while consuming 5 mA current from a 1.2 V supply, has efficiency of around 16.7%:

$$E_{PA} = \frac{10^{-3}}{5 \times 10^{-3} \times 1.2} \times 100 \cong 16.7\%$$

The PA efficiency defined by equation (E.2) is also known as the drain efficiency.

There are several factors that contribute to efficiency of the PA. For example, in Figure E.9, the Q of the inductor itself has a considerable impact on the PA efficiency. Generally speaking, the higher the Q of an inductor is, the better the PA efficiency will be. Therefore, if the inductor in Figure E.9 is an on-chip inductor, the Q will be low and the PA efficiency will be relatively low as well. The inductor can be implemented using the bond-wires (from the silicon die to the package substrate) or an external inductor can be used to improve the PA efficiency.

Power amplifier topologies are designated as classes such as A, B, C, D, E, F, G, H, and S. This classification is based on the shape of the current waveform in the PA. In class A, for example, the DC current of the PA is high enough that transistor, which is responsible to provide the gain, stays in active region all the time. Class A is a very linear PA, but the efficiency is low. In class A, the output of the amplifier is a full waveform. In class B, in contrast, the gate bias point of transistor M1 in Figure E.9 is selected at the threshold voltage. Therefore, the transistor M1 can only amplify the input signal when the V_{in} is more than the M1 threshold voltage. The efficiency of class B is better than class A and the averaged quiescent current is lower in class B. Class AB amplifier is an amplifier that is half-way between class A and class B. The amplifier in Figure E.9 normally can be biased as a class AB. For an IEEE 802.15.4 transceiver, class B and class AB are commonly used to implement the integrated PA.

In class E power amplifiers, the transistor M2 in Figure E.9 is not present and the transistor M1 is used as a switch. The transistor M1 is turned on for half of the period of the signal and is turned off for the reminder of the period. The inductor L in Figure E.9 must be large enough for class E amplifiers to ensure continuous flow of the current. The matching circuits will filter out the additional harmonics and pass only the fundamental frequency of the input signal. Class E amplifiers are nonlinear and highly efficient. Other PA classes are normally used only for high power PAs and may not be the best option for low power IEEE 802.15.4 transceivers. Reference [5] provides additional information regarding other classes of power amplifiers.

Linear amplifier is required when the signal contains both amplitude and phase information. In OQPSK and BPSK modulation techniques, the information is only in the phase of the signal. An introduction to OQPSK and BPSK modulations is provided in chapter 4. The OQPSK modulated signal delivered to the power amplifier has an almost constant envelope. Since there is no information in the amplitude of a OQPSK or BPSK signal, the PA can be designed as a nonlinear amplifier, which is more power efficient than a linear amplifier.

The nonlinear gain of a PA is represented by a nonlinear amplitude gain (Amplitude to Amplitude conversion or AM-AM) and a nonlinear phase gain (Amplitude-to-Phase conversion or AM-PM). The nonlinear phase gain is the signal phase degradations due to variation in the amplitude of the input signal. AM-to-PM conversion is usually defined as the change in output phase for a 1-dB increment in the PA input signal power. The AM-PM is expressed in degrees-per-dB ($°/dB$). An ideal amplifier has AM-PM of zero degrees-per-dB. The presence of AM-PM in a power amplifier implies that even in a constant envelope modulated signal, the nonlinearity of the amplifier can degrade the signal phase. One of the reasons for AM-PM conversion in a power amplifier is voltage dependent devices in the circuit. For example, the value of a capacitors may vary based the voltage applied to the capacitor. A capacitor adds phase shifts to the signal applied to the capacitor and therefore, the voltage variations will result in phase variations in the signal. Adding a time varying noise to the signal phase will degrade the signal error vector magnitude (EVM). The EVM is discussed in chapter 4.

The performance of an integrated PA depends highly on its output impedances. By varying the output impedance and measuring the PA performance, it is possible to find a sweet-spot in which the PA performance is optimum. The experiment of varying (pulling) the output impedance of a PA and measuring its performance is known as *Load-Pull Test*. This test is normally automated because it requires several measurements. The information from a PA load-pull test allows a PA designer to optimize the PA gain, linearity, and efficiency.

E.4   Frequency Generation

Frequency generation is one of the challenging tasks in a transceiver. A transceiver may not only require a precise reference frequency signal, but also may need low-power low-accuracy clocks as well. The high accuracy reference signals are generated using crystal oscillators. A basic introduction to crystal oscillators is provided in chapter 4. The synthesizer block in a transceiver generates a signal with variable frequency. The main block inside the synthesizer is the Voltage Controlled Oscillator (VCO). The VCO generates a tunable RF signal. The VCO output signal is used by several blocks in a transceiver. For example, the mixers in the receiver multiply the VCO output signal by the input RF signal to down convert the RF signal to the baseband. The operation and performance requirements of a VCO are reviewed here using the simplified VCO schematic example shown in Figure E.10a. The VCO in Figure E.10a is only one of several VCO circuit topologies, but the discussion in this section is applicable to most LC-VCO configurations.

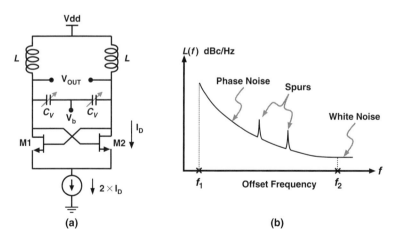

Figure E.10: (a) Simplified Topology of an LC-VCO (b) Generic VCO Phase Noise Spectral Density Versus Offset Frequency

The VCO in Figure E.10a is unstable in nature and starts oscillating if the transistors M1 and M2 have sufficiently high g_m values [3]. The value of the varactors (C_V) and the inductors (L) determine the oscillation frequency of this VCO:

$$f_{OSC} \cong \frac{1}{2\pi \times \sqrt{L \times (C_V + C_P)}} \text{(Hz)} \tag{E.3}$$

where C_P is the total parasitic capacitance on each side the VCO output. An ideal VCO generates an ideal tone, which is a sinusoid signal with bandwidth of equal to zero. However, the PSD of any VCO not only contains the sinusoid tone, but also includes undesired noise around the ideal tone. Figure E.10b shows the VCO Power Spectral Density versus frequency. It is common to plot the VCO output in dBc versus the offset frequency from the oscillation frequency. In Figure E.10b, the $L(f)$ is the power spectrum of the phase noise in dBc/Hz plotted versus offset frequency. The VCO phase noise is measured from offset frequency of f_1 to offset frequency of f_2. The offset frequency of f_1 must be as low as possible because the phase noise energy is significantly stronger at lower frequencies. The value of f_1 is determined by the practical limitations of the phase noise measurement instruments. Figure E.10b shows that the VCO output signal not only contains phase noise, but also undesired spurs may appear in the VCO PSD.

The VCO phase noise in Figure E.10b corresponds to an equivalent phase error. The root-mean-square (rms) of this phase error ($\Delta\varphi$) can be determined from the equation below:

$$\Delta\varphi = \sqrt{2 \times \int_{f_1}^{f_2} 10^{\frac{L(f)}{10}} \, df} \tag{E.4}$$

where the $\Delta\varphi$ is in radians, the $L(f)$ is in dBc/Hz, and f is the offset frequency in Herz. The rms value of the phase error ($\Delta\varphi$) is very valuable and can be used to determine some of the key performance metrics of the VCO. For instance, the relationship between the VCO phase noise and the EVM caused by the VCO can be approximated from the equation below:

$$\text{EVM} (\%) \approx 100 \times \Delta\varphi \quad (\Delta\varphi \text{ is in Radians}) \tag{E.5}$$

For example, if the rms value of the phase error ($\Delta\varphi$) is 6 degrees (0.104 Radians), then the EVM of the signal generated by VCO is 10.4%.

Another performance metric that can be extracted from the $\Delta\varphi$ is the rms value of the signal jitter (ΔT). The jitter (in seconds) is the undesired variations of the signal period.

$$\Delta T = \frac{\Delta \varphi}{f_0 \times 2\pi} \tag{E.6}$$

where the ΔT is the signal jitter in seconds, the $\Delta \varphi$ is the phase error rms value in radians, and f_0 is the oscillation frequency in Hertz. For instance, if the rms value of the phase error ($\Delta \varphi$) is 1 degree (0.017 Radians) in a 100 MHz signal, then the jitter of the signal will be 28 picoseconds. The ideal period of a 100 MHz signal is 10 ns, but due to 28 ps of jitter, the period of this signal varies from 9.986 ns to 10.014 ns.

If a signal $s(t)$ with frequency of f_{RF} is multiplied by an ideal tone with frequency of f_{RF}- f_{IF}, the signal $s(t)$ will be down converted to f_{IF} frequency and the PSD of the signal $s(t)$ remains unchanged. Figure E.11 shows a simple way of calculating the estimated phase noise requirement for a VCO block. The desired signal is at frequency of f_{RF}. This simple method is based on two simplifying assumptions. First of all, when the VCO signal is multiplied by the desired signal, the VCO signal is assumed to be an ideal tone without any phase noise. The interferer is also assumed to be an ideal tone with frequency of $f_{RF} + \Delta f$. The desired signal in Figure E.11 is down converted to f_{IF} frequency. The interferer (an ideal tone) down converts the VCO signal to $f_{IF} + \Delta f$. Part of the VCO signal after the down conversion will be in the same frequency band as the desired signal. The Signal to Noise Ratio (SNR) is defined as the ratio of the signal energy to the

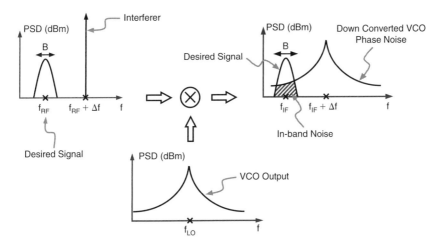

Figure E.11: A Simple Method of Calculating the Effect of VCO Phase Noise on the Desired Signal SNR after Down Conversion

energy of the noise within the frequency band of interest. The averaged VCO phase noise at Δf frequency offset (n_{VCO} (dBc/Hz)) can be calculated from the following equation:

$$n_{VCO} = P_D - P_I - SNR - 10 \times \log_{10}(B) \tag{E.7}$$

where,

B (Hz)   = The desired signal bandwidth

P_D (dBm) = The desired signal power (within the bandwidth "B")

P_I (dBm) = The interferer signal power (within the bandwidth "B")

Δf (Hz)   = The frequency difference between the center of the desired signal and the center of the interfering signal

SNR (dB) = The desired SNR

For example, consider a transceiver operating at 2.4 GHz with signal bandwidth of 2 MHz. There is an interfering signal, 55 dB stronger than the desired signal, at 5 MHz away from the desired signal. In order to ensure at least 15 dB of Signal to Noise Ratio, the VCO phase noise must be -114 dBc/Hz at 5 MHz away from the carrier frequency:

$$P_D - P_I = -55dB$$

$$
\begin{aligned}
n_{VCO} &= -55 - 15 - 10 \times \log_{10}(2 \times 10^6) \\
&= -114dBc/Hz \text{ (averaged over 2MHz bandwidth)}
\end{aligned}
$$

The desired SNR is the SNR of the signal after down conversion by the mixer. This SNR does not account for any de-spreading performed by the receiver. The positive effect of signal de-spreading on SNR is discussed in chapter 4. Calculating the required n_{VCO} based on the relative amplitude of the interferer signal is only one of the ways to determine the VCO phase noise specification.

Generally, the phase noise of an LC-tank VCO can be improved by increasing the Q of the LC tank. Increasing the LC tank Q helps reducing the power consumption of the VCO as well. External inductors and bond-wires have better Q compare to on-chip inductors. One way to improve the Q of the inductor is using external inductors and including the bond-wires as part of the total inductance. The phase noise requirement of IEEE 802.15.4 transceivers is not very tight and therefore, the IC manufacturers avoid using an external inductor to save area and cost.

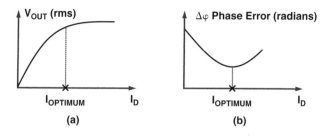

Figure E.12: (a) VCO Output Voltage Versus I$_D$ (b) VCO Phase Error Versus I$_D$

Figure E.12 shows the relationships between the VCO phase noise, I$_D$, and VCO output voltage (V$_{OUT}$). In the linear region (Figure E.12a), increasing I$_D$ increases the VCO output voltage. Furthermore, the phase noise of the VCO is reduced as the current I$_D$ is increased. Therefore, in the linear region, the VCO noise performance can be improved at the cost of increasing the power dissipation. The VCO output voltage increases linearly with I$_D$ until the VCO enters the nonlinear region. In the nonlinear region, increasing the I$_D$ degrades the VCO phase noise. A VCO is typically designed to operate at sweet spot (I$_{OPTIMUM}$) in Figure E.12b.

Dividing the signal frequency can improve the phase noise of the VCO output signal as long as the noise contribution of the frequency divider block is negligible. Multiplying the signal frequency, on the other hand, degrades the signal phase noise.

The VCO does not have any input signal. In a receiver chain, if there is a strong interferer at the LNA input, the interferer may leak to the VCO block through mixer. The VCO frequency of operation can be affected if a signal is injected to the VCO block. Generally, the VCO frequency of operation changes toward the frequency of the injected signal. This is known as *injection pulling* issue. Increasing the reverse isolation between the VCO and the mixer can reduce the injection pulling.

The injection pulling can be caused by the Power Amplifier (PA) as well. The PA output is a strong signal and may leak back to the VCO if the reverse isolation between the PA and the VCO is not sufficient. If the PA output frequency is close to the VCO frequency, the VCO frequency of operation may be pulled toward the PA frequency. In the direct conversion transmitters, where the VCO frequency is the same as the PA output signal frequency, the PA leakage to the VCO can disturb the VCO output PSD and degrade the VCO phase noise and its spectral purity. In this way, the VCO pulling translates to the transmitted signal EVM degradation.

One of the methods that can be used to reduce the VCO pulling is operating the VCO at twice the channel frequency. In this way, the VCO operation frequency is not the same as the PA output signal frequency. However, if the PA is very nonlinear, the second order harmonic of the PA can contribute to VCO pulling. The VCO pulling will be reduced if the PA output frequency and the frequency of its harmonics are sufficiently higher or lower than the VCO oscillation frequency.

A dual-conversion transmitter architecture with proper IF frequency can reduce the effect of VCO pulling. Operating an LC-VCO at twice the channel frequency can also reduce the VCO area because the size of the VCO inductor is reduced as the frequency of operation is increased. The coupling between the PA and the VCO can be via the shared power supply, the common ground lines, or the substrate. If the VCO pulling is due to the poor reverse isolation between the PA and the VCO, a buffer stage can be used to improve the reverse isolation between these two blocks.

The supply voltage variation in Figure E.10a translates to variations in the voltages across the varactors. Therefore, the ripple on the supply voltage can change the VCO frequency of operation. This is referred to as *supply pushing* problem. In sensitive RF transceivers, it is recommended to have a separate voltage regulator dedicated to the VCO to minimize the supply voltage ripple.

The capacitance value of a varactor is a function of the voltage applied to its terminals. Therefore, V_b in Figure E.10a controls the oscillation frequency of the VCO. The VCO tuning range is defined as the frequency ranges the VCO can cover by adjusting the value of C_V. The tuning sensitivity is defined as the ratio of the VCO frequency variation to the variations of the control voltage V_b. The tuning sensitivity is represented by K_V:

$$f_{OSC} = f_0 + K_V \times V_b \qquad (E.8)$$

$$\Delta f_{OSC} = K_V \times \Delta V_b \qquad (E.9)$$

where Δf_{OSC} is the oscillation frequency variation, ΔV_b is the control voltage variations, and f_0 is the VCO oscillation frequency when V_b is equal to zero. The control voltage V_b may contain noise and Equation E.9 shows that the sensitivity of the VCO frequency to the noise is proportional to K_V. Reducing K_V reduces the tuning range, but improves the VCO phase noise.

The LC-VCO is only one of circuit topologies that can be used to generate a tunable RF signal. A ring oscillator, for example, is another type of oscillator commonly used in integrated circuits. A ring oscillator consists of several delay stages that form a loop. Ring oscillators do not have any inductor and the oscillation frequency is controlled by adjusting the time delay of each stage of the ring. The phase noise and power dissipation of ring oscillators are generally worse than comparable LC-VCOs. The ring oscillators require smaller area and are less susceptible to pulling when compared to the LC oscillators.

The VCO circuit is normally part of a frequency synthesizer block. The frequency synthesizer uses a reliable and accurate reference frequency to generate various frequencies required in a transceiver. The VCO frequency of operation is typically controlled by a phase locked loop (PLL). In this section, operation mechanism of a typical PLL block is reviewed. The references [14] and [16] provide valuable information regarding synthesizer design and trade-offs.

Figure E.13a is a conventional charge-pump based PLL. The reference frequency is a fixed and accurate signal generated by a crystal oscillator. The VCO output frequency is divided by the frequency divider and is compared to the reference frequency. The difference between the VCO frequency (after division) and the reference frequency is the frequency error (Δf):

$$\Delta f = f_{REF} - f_{DIV} \tag{E.10}$$

The output signal of the Phase/Frequency Detector (PFD) block contains spurs and can create undesired frequency modulation in the VCO output. The loop low-pass filter

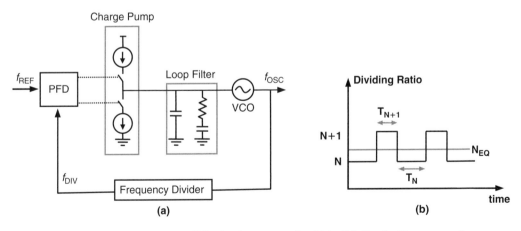

Figure E.13: (a) Simplified Diagram of a PLL (b) Basic Concept of Fractional-N Division

suppresses the spurs generated by the PFD. The clock feedthrough also creates spurious signals. Lowering the bandwidth of the loop filter reduces the effect of the spurs, but degrades the PLL transient response. In other words, suppressing the spurs comes at the cost of increasing the PLL switching time between two frequencies.

In a charge-pump based PLL, the acquisition time is directly proportional to the initial frequency difference (Δf) and inversely proportional to the loop bandwidth. The charge pump creates a constant current pulse. The duration of this pulse is proportional to the phase difference between the f_{REF} and f_{DIV}. The polarity of the current is the same as the polarity of Δf. The loop filter converts this current into a voltage that controls the VCO frequency.

If the frequency divider in Figure E.13a divides the frequency by an integer number, the PLL is referred to as an integer PLL. In an integer PLL, the reference frequency cannot exceed the channel spacing, which can be a limiting factor. If the frequency divider can divide the VCO frequency to a factional number between N and N + 1, then the PLL in Figure E.13a is a fractional-N PLL. The common method used to create a fractional divider is shown in Figure E.13b. The divider divides the signal to either N or N + 1 and the ratio of T_{N+1} to T_N determines the equivalent average dividing ratio (N_{EQ}):

$$N_{EQ} = N + \frac{T_{N+1}}{T_N + T_{N+1}} \tag{E.11}$$

In a fractional-N PLL, the reference frequency can be much larger than the channel spacing.

In recent years, digitally intensive PLL designs have been explored by many researchers. An all-digital PLL is presented in [14], where in contrast to the traditional PLL in Figure E.13a, there is no charge pump. The VCO frequency of operation is controlled by digital signals instead of a single analog voltage. The digital signals control a bank of varactors and digitally changing the equivalent capacitor. It is shown by [14] that the PLL noise, spurs, and switching time of an all digital PLL is superior to the traditional PLL of Figure E.13a.

One of the duties of the frequency generator is creating the in-phase and quadrature (I/Q) signals. The I and Q signals are two sinusoid signals with 90 degrees phase difference. The I/Q signals are used in the receiver path to demodulate an OQPSK signal. In an ideal scenario, the phase difference between the I and Q signals are exactly 90 degrees and the gain in the I and Q channels are the same. The deviation from these ideal scenarios is inevitable and results in signal EVM degradation.

A useful way of determining the source of the EVM in a signal is plotting the measured signal constellation and comparing it with the ideal scenario. Figure E.14a shows the

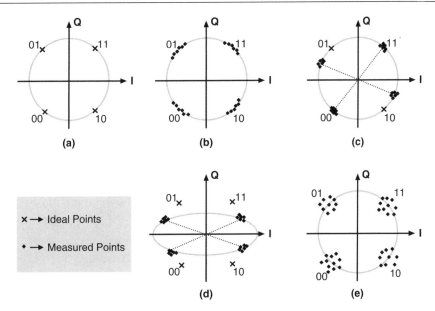

Figure E.14: (a) The Ideal Constellation Point in OQPSK Modulation (b) Effect of Phase Noise (c) Effect of I/Q Phase Imbalance (d) Effect of I/Q Channels Gain Imbalance (e) Effect of WGN

ideal constellation points for OQPSK modulation. In Figure E.14b, for instance, only phase error is present and there is no amplitude error. Also, the I and Q signals have approximately 90 degrees phase difference. The measured signal constellation points in Figure E.14b deviate from the ideal points due to the VCO phase noise and no other source of error is present in Figure E.14b. From Figure E.14c, it can be concluded that the phase difference between I and Q signals is not 90 degrees and the main source of EVM in Figure E.14c is I/Q phase imbalance. If there is no phase imbalance between the I and Q signals, but the gain in the I channel and Q channel are not the same, the constellation points will be stretched in one direction. Figure E.14d shows the effect of I/Q channels gain imbalance. Finally, in Figure E.14e, the main source of EVM is white Gaussian noise (WGN) and each constellation point is spread randomly around its ideal point.

E.5   Power Management

A transceiver consists of several blocks with different characteristics. The RF and mixed signal blocks, for example, may require a very clean and low noise power supply. The digital blocks, on the other hand, may stay operational even if the supply voltage drops considerably from its nominal value. The required voltage level for these blocks may also be different. The analog blocks may prefer a large supply voltage to improve their dynamic range. The digital blocks can operate with a lower supply voltage compared to analog blocks, which allows the digital blocks to reduce their power dissipation. The System-on-Chip transceivers have a Power Management (PM) unit to provide proper voltage supplies to various blocks inside the transceiver. A PM unit typically consists of a controller unit and several voltage regulators. The controller unit manages the regulator settings and turns on and turns off the regulators as needed.

Generally speaking, the voltage regulators can be divided into two categories of switching regulators and linear regulators. In a switching regulator, instead of continuously drawing current from the battery, a switch connects and disconnects the battery periodically to an averaging circuit. The duty cycle of this switch will determine the output voltage. Considering that the battery is not continuously discharged, the average current drawn from the battery is reduced. The basic concept of a switching regulator is covered in chapter 6. The linear regulators, in contrast, simply scale down the voltage to a desired value, without using any switching circuit. Switching regulators have better power efficiency but may require large size off-chip passive components. A switching regulator is typically accompanied with a linear regulator. Each linear voltage regulator inside a transceiver must meet certain specifications. In the remainder of this section some of these requirements are briefly reviewed. The goal of a regulator designer is to meet the performance requirement while minimizing the area, current, and number of external passive components.

One of the basic requirements of a regulator is its sourcing capacity. The sourcing capacity is the maximum current a regulator can provide to the blocks connected to this regulator. The current consumption of the regulator itself must be as low as possible to reduce the overall current consumption of the block. In the low duty cycle applications, the leakage current of the regulator, when the regulator is turned off, must be kept extremely low because the total battery life of a low duty cycle node is mainly determined by its leakage current during the sleep mode rather than the maximum current consumed during the active mode of operation.

Figure E.15 shows a time-varying signal (V_{in}) is regulated down to V_{out}. One of the tasks of a linear voltage regulator is to suppress the input signal variations (ripples). The Power Supply Rejection Ratio (PSRR) is the ratio of the voltage ripples at the input of the regulator to the voltage ripples at the output of the voltage regulator:

$$PSRR = 20 \log \frac{\Delta V_{in}}{\Delta V_{out}} \text{ (dB)} \qquad (E.12)$$

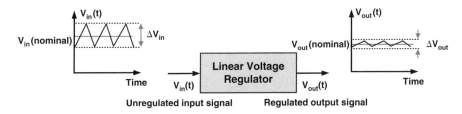

Figure E.15: Basic Concept of Voltage Regulation

For example, if the input signal has 200 mV ripple and the regulator reduces the ripple to 10 mV at the output, the regulator has PSRR of 26 dB. The Power Supply Rejection Ratio is also referred to as Power Supply Ripple-rejection Ratio. If the linear regulator is connected to a clean power supply such as a battery, the PSRR performance of the regulator is less critical. However, if the power supply has large ripples, the regulator must have sufficient PSRR to suppress the ripple and provide a clean regulated voltage to the sensitive RF and analog blocks.

The PSRR is typically measured over a range of frequencies. In a transceiver, there are certain frequencies that are more critical than others. The PSRR can be plotted versus frequency to analyze the regulator performance at frequencies of interest.

The active devices inside a regulator create noise (e.g., thermal and flicker noises). The noise at the output of the regulator is directly transferred to the block that is supplied by this regulator. This noise is a function of the frequency and the noise is normally higher at lower frequencies. Since different blocks inside a transceiver have different levels of tolerance for the noise injected by the regulators, it is recommended to separate the regulators of noise sensitive blocks (e.g., RF blocks) from the regulators of the block that are more noise tolerant (e.g., digital blocks).

E.6   Microcontrollers

Some of the transceivers developed for short-range wireless networking come with integrated microcontrollers. The microcontrollers used in battery-powered low-duty cycle applications typically support power saving modes of operations. The current consumption of the microcontroller, when operating at maximum clock frequency, can be comparable with the current consumed by the radio section of the transceiver. Therefore, in battery powered nodes, it is generally recommended to avoid turning on the radio and the microcontroller (at the maximum clock frequency) concurrently. This would help reduce the peak of the current extracted from the battery and helps improve the battery life.

The microcontrollers are classified based on the size of their data bus. For example, a microcontroller with 8-bit data bus is referred to as an 8-bit microcontroller. The size of the microcontroller address bus determines the maximum internal memory it can support. Most microcontrollers come with debugging capabilities and support various peripherals. The following are examples of microcontrollers that can be found in IEEE 802.15.4 System-on-Chip solutions.

E.6.1 ARM

The instruction set is the hardware language used by the software to operate a processor. Careful reduction of the instruction set allows optimizing the microprocessor design and can increase the microcontroller performance. The processors that optimize their performance by reducing their instruction set are commonly referred to as the Reduced Instruction Set Computers (RISC). The Advanced RISC Machines (ARM) microcontrollers are popular 32-bits RISC processors developed by ARM Ltd [1]. ARM Ltd. mainly licenses the processor architecture to other microcontroller manufacturers (e.g., Freescale semiconductor). ARM microcontrollers are known for their high performance and low power consumption. The ARM7 and ARM11 microcontrollers, for instance, have been widely used in battery powered mobile devices.

E.6.2 HC(S)08

The HC08 and HCS08 are low-power general purpose 8-bit microcontrollers from Freescale Semiconductor [6]. The clock frequency in HCS08 family can be up to 40 MHz. Similar to many other microcontrollers, HCS08 support various power saving operation modes. The low power wait mode feature of this microcontroller makes it a suitable choice for low duty-cycle short-range wireless networking applications.

E.6.3 8051

The 8051 is an 8-bit microcontroller (i.e., the data bus is 8 bits). The address bus in 8051 is 16 bits and therefore can support up to 64 KB of internal memory. The 8051 supports power saving modes, which is essential for low duty cycle battery powered devices. There are several manufacturers that develop 8051-compatible processors. These enhanced 8051 microcontrollers not only have additional features (e.g., brown-out detection), but also have improved the maximum clock rate of the 8051 core.

E.7 Interfaces

The interfaces allow a transceiver to communicate with other components outside the transceiver IC. One of the common interfaces available in System-on-Chip (SoC) solutions is General Purpose Input Output (GPIO). A GPIO pin can be configured as either input or output depending on the application. A keyboard interface would be beneficial if a transceiver is connected to a keyboard panel. An on-chip Analog to Digital Converter (ADC), discussed in chapter 4, can facilitate interfacing with sensors by converting the output of sensors from analog to digital. In this section, examples of interfaces commonly available in short-range wireless networking transceivers are briefly reviewed.

E.7.1 SPI

The Serial Peripheral Interface (SPI) is one of the widely used methods of establishing serial communication between two or more devices [15]. SPI uses master/slave mechanism to control the flow of the serial data (Figure E.16). There are four signals in the SPI bus:

- Master Output, Slave Input (MOSI)

- Master Input, Slave Output (MISO)

- Serial Clock (SCLK)

- Slave Select (SS)

The MOSI is used to transfer data from master to slave. The MISO transfers the data from slave to master. The SCLK synchronizes the serial interface communications. A slave can communicate with the master when its SS line is held low (or there is a falling edge from high to low). Every time the master sends a bit to slave through the MOSI, the slave sends a bit to master through MISO. In this way, the SPI interface provides full duplex communication because the data is transferred between the master and slave concurrently.

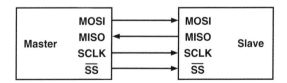

Figure E.16: SPI Master/Slave Communications

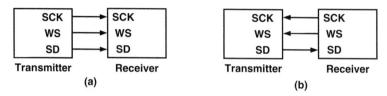

Figure E.17: I^{2}S Interface (a) Transmitter is Master (b) Receiver is Master

E.7.2 I^{2}S

The SPI discussed in the previous section was only one of several methods that can be used for serial communications between multiple chipsets. The Inter-IC Sound (I^{2}S) bus is an alternative way to establishing serial communication links between the various ICs [7]. An I^{2}S bus is intended to handle only digital audio data and the other signals (e.g., control signals) must be transferred separately. The I^{2}S bus, showed in Figure E.17, has three lines:

- Serial Clock (SCK)

- Word Select (WS)

- Serial Data (SD)

Since I^{2}S has only one serial line, the communication is not full duplex as in the SPI bus. In I^{2}S, the MSB is transmitted first. The WS determines which channel is being transmitted (left or right). The WS line changes one clock period before the MSB is transmitted to allow the transmitter to synchronize its timer, and the receiver will have a chance to store the previous word.

E.7.3 JTAG Boundary Scan Interface

An integrated circuit used in short-range wireless networking may contain RF, analog, digital, microcontroller, and memory on the same silicon die. The complexity of these ICs requires a standard method of testing and debugging. The IEEE released a standard entitled "Standard Test Access Point and Boundary Architecture" (IEEE 1149.1) to address this need [8]. The content of the standard came from Joint Test Action Group (JTAG). The JTAG was formed by a number of leading silicon manufacturers. A silicon manufacturer may optionally support this standard method of testing and provide the appropriate interfaces. The test interface is referred to as JTAG boundary scan interface or simply JTAG interface [9].

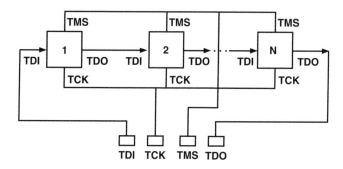

Figure E.18: JTAG Boundary Scan Interface

A JTAG interface allows testing interconnects of sub-blocks inside an IC without using physical probes. Figure E.18 shows the JTAG interface. The Test-Data-In (TDI) is used to provide a serial test data to a sub-block. The serial output of a sub-block is identified by Test-Data-Out (TDO). Each sub-block has a Test-Clock (TCK). The JTAG uses state-machines. The Test-Mode Select (TMS) allows changing the state of the state machine. The JTAG may optionally provide a reset pin (TRST) that can be used to reset the state of the state machine.

E.7.4 Nexus

The Nexus 5001™ Forum Standard for a Global Embedded Processor Debug Interface (or simply Nexus) is a comprehensive test and debugging standard for embedded processors [11]. Nexus is adopted by IEEE as IEEE-ISTO 5001™-2003. Nexus allows real time debugging and calibration. The JTAG interface can be found in many devices. Therefore, the Nexus standard defines an extensible Auxiliary Port (AUX) that may either be used with the JTAG interface or as a stand-alone development port. Further information regarding Nexus interface can be found in [11].

E.8 Packaging

In this section, some of the popular packaging methods used in low power short-range wireless transceivers are reviewed. Manufacturers always try to minimize the size and cost of the packages and improve the thermal, mechanical, and electrical performance of the package. The thermal performance of the package is a function of the package type, its geometry, materials, and the environment. The size of the die itself is a factor when determining the minimum possible package size. A package can support only a limited number of pins and sometimes it might be necessary to increase the size of the package or use a different packaging method to ensure the package can support the desired number of Input/Output (I/O) connections. Some of the high pin-count packages may require special setups for testing, and therefore, these packages may be more expensive to test and evaluate. Therefore, when selecting a package, not only the initial packaging cost and size must be considered, but also the cost of testing should be taken into account.

The Restriction of Hazardous Substances (RoHS) Directive of Europe restricts the use of certain hazardous materials [13]. Lead, Mercury, and Cadmium are examples of toxic materials prohibited by RoHS Directive. Until recently, lead could be found in many packaging solutions. Nowadays, many manufacturers offer RoHS compliant packaging options.

E.8.1   QFP and QFN Packages

The Quad Flat Package (QFP), shown in Figure E.19a, is one of most basic packaging options available. The leads are extended from each side of the package. The QFP is not new and has been used for decades. The QFP is relatively low cost and can be used in low to medium pin count applications. If the number of pins is high, other types of packaging including Ball Grid Array (BGA) can be used instead.

The Quad Flat No-lead (QFN) package, shown in Figure E.19b, does not have leads extending from each side. Instead, metal connections are provided under the package around the perimeter of the package. Figure E.19b shows a large ground flag under the

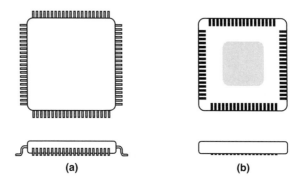

(a)   (b)

Figure E.19: (a) QFP and (b) QFN Packages

package, which not only provides ground connection, but also may help with thermal performance of the package.

E.8.2 BGA and LGA Packages

The Ball Grid Array (BGA) package, shown in Figure E.20, is advantageous when the pin count is high. In contrast to QFP, the BGA package can use the entire area of the bottom of the package for Input/Output (I/O) connections. In this way, using a BGA package can help reduce the package size in pin-limited scenarios. In Figure E.20, the die is placed on top of a substrate. This substrate can have multiple metal layers to route all the connections. The die can be encapsulated using a mold compound. The solder balls under the package simplify placement and handling of the package. In order to solder a BGA package to a board, the package is simply put in place and the heat is used to melt the solder balls and create interconnects between the package and the board. In BGA packaging, the lengths of electrical connections are shorter than QFP, which reduces the parasitic inductances and capacitances. The mechanical performance of the BGA is better than QFP because the BGA does not have fragile leads.

The BGA package does not have leads and therefore thermal expansion and contraction can put great pressure of the solder interconnects. Therefore the mechanical and thermal characteristics of the PCB should be matched to those of the BGA. In BGA, the solder balls may become damaged or disconnected due to shipping. The Land Grid Array (LGA) package is a BGA package without the solder balls. The LGA has metal pads in place of solder balls. The LGA solder interconnects are formed by applying solder paste directly to the board. Testing and evaluation of BGA and LGA packages may be more expensive than QFN and QFP packages.

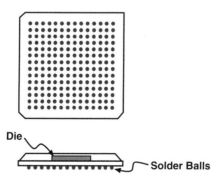

Figure E.20: Ball Grid Array (BGA) Package

References

[1] ARM Microcontrollers, available at http://www.ARM.com

[2] Ball Grid Array (BGA) Package, Intel® Packaging Databook, Chapter 14, Available at http://www.intel.com

[3] A. D. Berny, et al., "A 1.8 GHz LC VCO With 1.3-GHz Tuning Range and Digital Amplitude Calibration," *IEEE JSSC*, April 2005, pp. 909–917.

[4] A. Cavallini, et al., "Chip-Level Differential Encoding/Detection of Spread-Spectrum Signals for CDMA Radio Transmission over Fading Channels," *IEEE Transactions on Communications*, Vol. 45, No. 4, April 1997.

[5] A. Grebennikov, "*RF and Microwave Power Amplifier Design*," McGraw-Hill Professional, August 25, 2004.

[6] HCS08 microcontroller families, available at http://www.freescale.com

[7] Inter-IC Sound (I^2S) Bus Specifications, available at http://www.nxp.com

[8] IEEE Standard Association, available at http://www.ieee.org

[9] JTAG Boundary Scan Interface, available at http://www.jtag.com

[10] Land Grid Array (LGA) Package, Application Note AN2920, available at http://www.freescale.com

[11] Nexus Forum, available at http://www.nexus5001.org

[12] B. Razavi, "*RF Microelectronics*," Prentice Hall International, 1998.

[13] The Restriction of Hazardous Substances (RoHS) directive, available at http://www.RoHS.eu

[14] R. B. Staszewski and P. T. Balsara, "*All-Digital Frequency Synthesizer in Deep-Submicron CMOS*," John Wiley & Sons, 2006.

[15] SPI Application Notes, available at http://www.freescale.com

[16] C. S. Vaucher, "*Architectures for RF Frequency Synthesizers*," Kluwer Academic Publishers, 2002.

[17] X. Yang, et al., "A Digitally Controlled Constant Envelope Phase-Shift Modulator for Low Power Broad-Band Wireless Applications," *IEEE Transaction on Microwave Theory and Techniques*, Vol. 54, No. 1, Jan 2006, pp. 96–105.

Glossary

Term	Definition
Application objects	Application objects are developed by manufacturers, and that is where a device is customized for various applications. There can be up to 240 application objects in a single device.
Association	Establishing membership for a device in a wireless network.
Beacon-enabled WPAN	A WPAN in which all coordinators transmit beacon frames.
Binding	The task of creating logical links between the applications that are related.
Broadcast	A broadcast message is intended for any device that receives the message.
Broadcast jitter	The random wait period during which a device waits before each broadcast.
Broadcast transaction record (BTR)	The record that the ZigBee coordinator and ZigBee routers maintain of all the messages that they broadcast.
Broadcast transaction table	A table that contains broadcast transaction records.
Contention access period (CAP)	During CAP, any device that wants to transmit must compete for channel access using a CSMA-CA mechanism.
Coordinator	A full-function device (FFD) that is capable of relaying messages.
Encryption	Modifying a message into a new form in a way such that only the expected recipients can recover the original message.
Endpoint address	Each application object has a unique endpoint address (endpoint 1 to endpoint 240). The endpoint address of zero is used for the ZDO.

Data authentication	Verifying the true source of the information in the message and ensuring that the message has not been modified in transit.
Device description	Part of an application profile that provides information regarding the device itself.
Disassociation	A procedure that an associated device uses to notify the coordinator that the device intends to leave the network.
Full-function device (FFD)	An FFD is capable of performing all the duties described in the IEEE 802.15.4 standard and can accept any role in the network.
Indirect addressing	In indirect addressing, a device with limited resources that is bound with other devices in a network can communicate without knowing the address of the desired destination.
Key establishment	Establishing a security key for two or more devices.
Link key	A security key that is shared between only two devices.
Mesh topology	In mesh topology, in contrast to the tree topology, there are no hierarchical relationships. Any device in a mesh topology is allowed to attempt to contact any other device. This helps a mesh network have a self-healing capability.
Neighbor table	A ZigBee device uses a neighbor table to keep the information regarding the nearby devices.
Nonce	The nonce is a 13-octet string constructed using the security control, the frame counter, and the source address fields of auxiliary header.
Octet	An octet is equal to one byte (8 bits).
Orphaned device	A device that has lost its connection with its associated coordinator.
Personal area network (PAN) coordinator	The principal controller of a PAN.
Personal operating space (POS)	The spherical region that surrounds a wireless device and has a radius of 10 meters (33 feet).
Profile	A ZigBee profile is a collection of device descriptions
Radio sphere of influence	The region around the device in which its radio can successfully communicate with other radios.
Reduced-function device (RFD)	An RFD has fewer capabilities than an FFD and cannot act as a coordinator.

Route discovery	Finding a route that allows relay of messages from the source device to the destination device.
Route discovery table	The route discovery table contains the path costs, the address of the device that requested the route (source device), and the address of the last device that relayed the request to the current device.
Security key	A binary value that is shared between the corresponding devices and helps unlock the information.
Self-healing	The ability of a network to overcome certain issues without human intervention. For example, if one of the routers stops functioning due to exhaustion of its battery, the network can select an alternative route.
Self-organizing network	In a self-organizing network, the nodes can detect the presence of each other and create a network without human intervention.
Service discovery	In service discovery, a device requests another device in the network to provide detailed information such as its profile identifier or its ZigBee descriptors.
Trust center	Designated device in any secure ZigBee network that distributes the link keys as well as the network key to other devices.
ZigBee-specific	A feature implemented using ZigBee NWK, APL, or security protocol layers. This feature is supported by the IEEE 802.15.4 standard.

Index